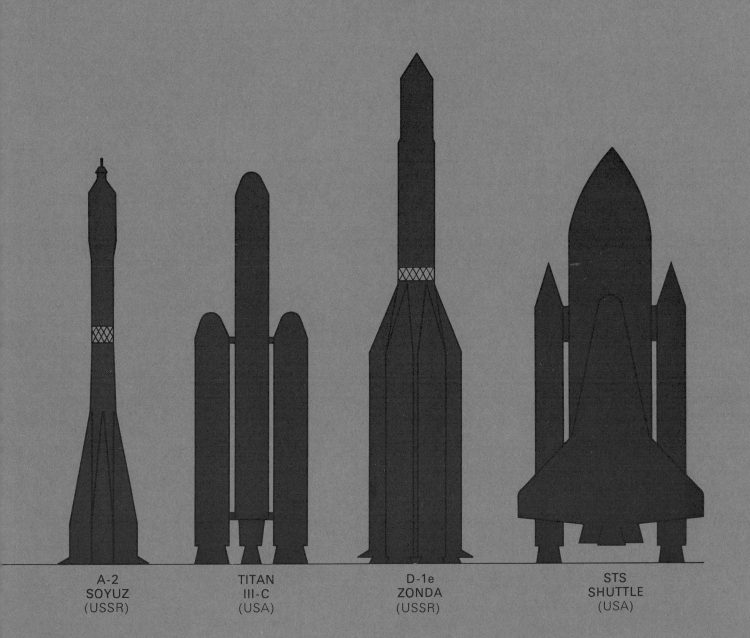

A-2
SOYUZ
(USSR)

TITAN
III-C
(USA)

D-1e
ZONDA
(USSR)

STS
SHUTTLE
(USA)

The Voyages of *Columbia*

The Voyages of
Columbia

The First True Spaceship

RICHARD S. LEWIS

New York   Columbia University Press   1984

Library of Congress Cataloging in Publication Data

Lewis, Richard S., 1916–
    The voyages of Columbia.

    Includes index.
    1. Columbia (Space shuttle)   I. Title.
TL795.5.L49   1984   629.44'1   82-1694
ISBN 0-231-05924-8

Columbia University Press
New York   Guildford, Surrey

Clothbound editions of Columbia University Press books are
Smyth-sewn and printed on permanent and durable
acid-free paper.

Contents

Prologue

The Space Shuttle marks the real beginning of space travel. It enables us to function in an extraterrestrial environment, a development of evolutionary scale analogous to the emergence of life from the sea in the far past. It is the most advanced artifact of 25 years of space technology. In that time, we have explored the Moon and become aware of its commercial and industrial potential. We have hurled instrumented probes to the planets—and one, *Pioneer 10,* beyond the known Solar System. Manned spaceflight has evolved through Vostok, Mercury, Voskhod, Gemini, Apollo, Soyuz, Salyut, and Skylab to the first reusable transport, *Columbia.*

The societal impact of this technology may not yet be apparent, but it portends changes in the outlook and vision of civilized people. Access to the interplanetary medium has conveyed a new perspective on the fragility of the Earth as the abode of life, and of its limits to growth.

The scientific results of planetary orbiters and probes and of astronomical observatories have altered ideas about the origin and evolution of the Solar System and about the nature of the universe. The citizen of the late twentieth century who is aware of discoveries in space will perceive many similarities among the terrestrial planets, and these may affect his world view. Close-up photos of Oceanus Procellarum on the Moon look like plowed fields in Kansas. Chryse Planitia on Mars resembles a stretch of Arizona desert. Such images enlarge our perception of our environment from the Earth to the Solar System. This transformation in perspective is as fundamental as that which dawned in the fifteenth century with the discovery of America.

It has been suggested that the invention of space travel promises our species immortality, or at least longevity, in that it enables us to tap the resources of matter and energy beyond the Earth and thus escape the limits to growth and ultimate survival imposed by the waning resources of one world. The key that unlocks this cosmic larder is a reliable and economically rational system of space transportation, one that ultimately will take us anywhere we want to go in the Solar System. With the development of the reusable Space Shuttle, we have taken the first, long step toward creating such a system.

It should be remembered that the Shuttle was proposed in 1969 as an element of an interplanetary transportation system that would enable the United States to land a crew on Mars in the 1980s. The system contemplated space stations in Earth and lunar orbits, orbital transfer vehicles, and lunar and planet landers. Our failure to develop the system beyond the Shuttle can be attributed in part to cost constraints resulting from the redefinition of national priorities that accompanied the social turmoil of the late 1960s and early 1970s. By the time the Shuttle was authorized in 1972, the urgency for developing manned spaceflight that had characterized the Apollo era was gone. The idea of going to Mars, which had seemed so appealing in the euphoria of the early lunar landings, became irrelevant to a government beset with social and racial tensions. The goad of technological competition in space with the Soviet Union that had galvanized Apollo had vanished with the landing of *Apollo 11* in Mare Tranquillitatis.

Compared with Apollo, which had had the highest priority, the Shuttle's priority was minimal. In order to get the project started at all, the National Aeronautics and Space Administration had minimized its cost and order of technical difficulty. As a result, the Shuttle was chronically underfunded and development was frequently held up by shortages of parts.

The nine-year period of the Shuttle's construction was punctuated by congressional criticism of the project's fiscal management, cost overruns, and repeated mechanical breakdowns. At times, the project was threatened with cancellation. This might have happened had it not been for the support of the Department of Defense, which persuaded Congress that the Shuttle would be an important adjunct to national defense.

From years of failure, frustration, and criticism, a reusable space transportation system has emerged for low Earth orbit. Two orbiters, *Columbia* and *Challenger*, are flying. One more, *Discovery*, is being prepared for flight, and the fourth, *Atlantis*, is in production.

The indifference and doubt that attended the development years have given way to an optimistic perception of this new vehicle as a means of opening up a vast new frontier to commercial and industrial enterprise as well as providing economical transport for satellites.

In a decade we have created the world's first fleet of true spaceships, an incomparable national resource. In light of its long travail, the accomplishment seems miraculous.

One of the great chapters in American history is the story of how that miracle came to pass. Still, it is only a beginning.

R.S.L.
Indialantic, Florida
January 8, 1984

The Voyages of *Columbia*

The Winged Spaceship

Edwards Air Force Base, California: It was cold before dawn in the high desert. I parked the rental Ford at the side of a silent hangar and sat there a few moments as the engine dieseled, coughed, and then died with a rattle. It was a long mile to runway 22-04 via an alkali road that could be distinguished from the surrounding waste of sagebrush and rusting automotive and aircraft parts by its lighter hue in the morning moonlight.

Runway 22-04, 15,000 feet long, was to occupy a special niche in the history of the space age this morning, October 26, 1977, as the scene of the last approach and landing test of a new spaceship, Space Shuttle–Orbiter 101. It was a hybrid, a product of technical miscegenation between an airplane and a spacecraft, and it was very large, the size of a commercial DC-9.

Boosted into orbit by rockets, the Shuttle-Orbiter was designed to reenter the atmosphere nose first and land like an airplane, or rather, a very heavy glider. It took some stretching of the imagination to visualize the return of the new spaceship from orbit. It would reenter the atmosphere over the Pacific Ocean southeast of Hawaii and glide, without power, 4,800 miles to a precision touchdown, which was planned initially on a 15,000-foot runway at the Kennedy Space Center on the east coast of Florida.

No machine had ever been designed to glide one-fifth of the way around the world to make a pinpoint landing—especially not one weighing 75 tons. Orbiter 101 was the first. Today's test would show whether it could make a precise landing on a controlled glide from about 17,000 feet. On the outcome depended the immediate future of American manned spaceflight and, in the longer range, the realization of a space agency–sponsored dream to develop the commercial and industrial potential of the solar system: orbital factories manufacturing superior pharmaceutical and electronic products in a gravity-free, hard vacuum environment; titanium and aluminum mines on the Moon; iron and heavy metals mines on

Artists' conception (circa 1977) of the final approach of the Shuttle-Orbiter to the three-mile runway at the Kennedy Space Center, Florida. To the right, off the starboard wing, is the Vehicle Assembly Building, where Shuttles are assembled for launch. (NASA)

Mars; human colonies on the Moon, on Mars, and in space; processing and assembly plants in space for the production of solar-electric power satellites to beam energy to Earth; a large space telescope that might reveal other planetary systems, the edge of the universe, and the beginning of time.

Technically, the dream could be realized; philosophically, it could be rationalized as another rung in the ladder of evolution. Life had begun in the sea, had evolved there and on land, and now was ready to expand into the cosmos. Only three-quarters of a century had elapsed since the Russian mathematics teacher and spaceflight visionary, Konstantin Tsiolkovsky, had written the rationale of the space age: "The Earth is the cradle of mankind, but one does not forever remain in the cradle."[1]

Now the link between dream and reality was the Space Shuttle. It was the basis of a space transportation system that, with the development of auxiliary vehicles, would enable man to cruise throughout the Solar System. The Shuttle itself was designed to travel only between the ground and orbits up to 600 miles high. Other vehicles, chemically and nuclear powered, would move crews and equipment to high orbits and to the Moon and planets.

The Shuttle was a system in itself. It comprised the 122-foot long space airplane, the orbiter, with a 78-foot wingspread; a huge propellant tank containing liquid hydrogen and liquid oxygen that fed the orbiter's main rocket engines; and two massive solid-fuel rocket boosters strapped to the sides of the big tank at launch. After the launch, the boosters would be dropped off at burnout and descend on parachutes to the sea, where they would be recovered for reuse. The tank would be jettisoned when the orbiter's rocket engines shut down, just before it entered orbit. It would not be recovered.

The Shuttle project had taken shape in 1969, before *Apollo 11* lifted off for the Moon. It was not only a technical advance over Apollo but an economic one as well. The first generation of American manned space vehicles were throwaways. They could be used only once and then, when recovered, placed on exhibit in some museum or state fair. The Shuttle-Orbiter was designed to be flown up to 100 times. It had to be designed that way if space travel was ever to become economically reasonable. In the context of a cold-war race to be first on the Moon, the cost of putting a crew in space with a quarter of a billion dollars' worth of expendable hardware might be justified; but once the race was won, the high cost of continuing manned spaceflight with disposable vehicles seemed intolerable. Hence the Shuttle, the first reusable spaceship.

For nearly ten months, Orbiter 101 had been undergoing a series of graduated approach-and-landing tests at the National Aeronautics and Space Administration's Dryden Flight Research Center, Edwards Air Force Base, in the Mojave Desert about 125 miles northeast of Los Angeles. The base sprawls over some 11,000 acres of the desert, and part of this desolate reservation is used by Dryden for testing advanced aircraft. It is an ancient sea bottom uplifted in an earlier epoch to 2,000 feet above sea level, surrounded by volcanic mountains, and pocked with dry lakes. The bed of one of the largest, Rogers Dry Lake, is the site of several packed-earth runways—one of them 7.5 miles long.

During the drive up there from Los Angeles International Airport, I turned to another sort of wasteland, AM radio, and thus learned that the principal VIP observing the test would be Prince Charles, the heir to the British throne. He had been touring the United States, and because of his enthusiasm for flying, NASA had taken him in tow and was showing him the marvels of its billion-dollar spaceflight establishment.

The fifth and final approach-and-landing test of the orbiter was to begin at 8 A.M. I reached the Edwards gate about 5:15 A.M. and, after a sequence of misinterpreted directions from white-gloved Air Police, wound up in the wrong parking place beside an empty hangar about 6 A.M. It was still dark, windy, and damn cold.

When the Ford had gasped its last, I climbed out and peered around the wilderness. Two other cars were parked alongside the hangar, having been directed (or rather misdirected) there, as I had been. A young woman was trying to read by flashlight a map that she had spread out on the trunk of her car. She said that she had risen from the side of her sleeping husband at 4 A.M. and had driven all the way from nearby Lancaster to cover the prince. She hoped fervently that he would show up to make the effort worth while. The other person was a brush-bearded photographer for a technical magazine. He was unloading equipment from a badly dented van, like one of those anonymous vehicles used by myriads of nomadic kids for housing as well as transportation on the highways of California.

At this point in time (a cliche which I am certain had its origin in those evasive news conferences NASA perpetrated during the adolescent phase of its development), Orbiter 101 was already securely bolted onto the back of its carrier aircraft, the Boeing 747 jumbo jet that would take it aloft and then release it for the glide test. NASA had yielded to the clamor of fans of the TV series *Star Trek* and dubbed the spaceship the *Enterprise,* after the starship of Captain Kirk, Mr. Spock, Lieutenant Uhura, et al. The concession to the "Trekkies" was logical, as Mr. Spock might say. They were the most visible element of the population outside of the aerospace industry that retained some enthusiasm for manned spaceflight after the end of Apollo and the landings on the Moon.

Unhappily, from the Trekkie viewpoint, the *Enterprise,* although the first orbiter to be manufactured and tested in the lower atmosphere, would not be the first of its generation in space. Orbiter 102, then being assembled at the Palmdale plant of Rockwell International, was slated to be first. The *Enterprise* itself would never reach orbit.

The currently abandoned Moon, still alive with nuclear-powered instruments transmitting geophysical data of declining interest to science, was setting over the Tehachapi Mountains—so that direction was west. The feature writer, the photographer, and I agreed that the runway must lie due south, toward the San Gabriel Mountains. At least, the road meandered in that direction. We began walking toward distant lights. The air was clear, chill, and aromatic with the faint smell of jet fuel exhaust. There was a distant roaring; NASA chase planes were warming up. Slowly, the stars dimmed and went out, like house lights just before the curtain rises.

A dark blue Air Force pickup truck rolled up and stopped long enough to allow us to climb over the tailgate into the cargo bay. The driver took us to the runway, discharged his passenger in the front seat and waited just long enough for us to jump out before speeding away. NASA's firm rule was: No vehicles to be parked within a half mile of the runway. The metal would interfere with the testing of the microwave scanning beam landing system. One day that system would land the orbiter automatically. But not today. Today, astronauts would land it.

Tea Party

At dawn, the horizon turned pink, like the skies of Mars. Magically, the desert became illuminated in cerise and gold. Amid the litter of rocks and rusting metal parts there appeared the mirage of a Mad Hatter's Tea Party. Three round tables, covered with white linen and set with gleaming silver, crystal, and red napkins, stood on a red carpet some 30 feet square that lay outstretched and lumpy on the sand and gravel.

White cane chairs with red plush seats were drawn up to the tables. Young women in white aprons tended a long serving platform where urns of coffee, tea, and hot chocolate were heating over the flickering blue flames of sterno cans. This Dali-esque scene was bounded by a rectangle of thick white cotton rope, supported by theater stanchions on flared bases. Beyond stretched an asphalt-concrete runway, from one misty infinity to another; beyond that, the desert, the mountains, the rosy morning sky.

The Dryden Flight Research Center was prepared to greet Charles Philip Arthur George, Prince of Wales. The Los Angeles caterer, specializing in weddings, confirmations, and bar mitzvahs, had set a table royal in the desert.

At the Lyndon B. Johnson Space Center in Houston, where Shuttle flights would be controlled, the 28-year-old prince had flown the orbiter simulator, an $18 million test chamber rigged like the orbiter's flight deck. In it a pilot could acquire the illusion of actually landing the vehicle on the cinematic desert scene projected below.

Along the south side of runway 22-04, the main paved runway of the Research Center, nearly a thousand men, women, and children had assembled by 7 A.M. Most were Air Force personnel and their families. They stood patiently behind a clothesline barrier strung on steel poles that extended for about a mile to warn people off the runway. Curiously, no one ventured to invade the desert Tea Party, although anyone could have stepped over the catenary sag of the ropes. It seemed to be as taboo to the hoi polloi as an Indian burial ground.

The *Enterprise* loomed up in the distance, mounted piggyback atop its 747 carrier as both were slowly towed toward the southwest end of the runway for the 8 A.M. takeoff. In silhouette against the misty sky, the stack suggested a construction of Leonardo da Vinci. Pilots at Edwards called it the world's biggest biplane.

Prince Charles arrived then with his escort in a long black Cadillac (rented), followed by a retinue of NASA officials and astronauts in a procession of lesser vehicles. The caravan crossed the runway and pulled up before the Tea Party scene, and the occupants moved smartly onto the compound like a troupe of actors who assemble on stage seemingly oblivious to the audience. News media still and television photographers swarmed around the compound.

The prince was a slender, medium-sized young man with a strong resemblance, several women journalists remarked, to his great-uncle, Edward VIII, of romantic memory. Some thought he bore the look of his great-great-great-grandmother, Queen Victoria. On the side of his mother, Elizabeth II, he was a direct descendant of George III. Media folk had checked that out. Some of them thought he was a better story than the Shuttle.

It is not remarkable that members of the House of Hanover, Saxe-Coburg, Wettin-Windsor appear familiar to Americans. They all look like people in a suburban PTA. It was said later that the future king of England asked good questions and seemed to be a quick study. Although Charles was a practicing airplane pilot, the orbiter was not like any airplane he had ever known.

For its part, royalty was surrounded by celebrity, and the twain got on well. Among Charles' hosts were David R. Scott, director of the Dryden Center, and Major General Thomas P. Stafford, commander of Edwards Air Force Base. History will remember Scott as the commander of the *Apollo 15* expedition to the Apennine Mountains on the Moon in 1971 and Stafford as commander of the American crew in the 1975 Apollo-Soyuz Test Project, the joint flight with the Russians that demonstrated detente in orbit. That had been Apollo's last hurrah.

The prince conversed with considerable interest and animation with astronaut Joe H. Engle, one of the orbiter's four test pilots, who had won astronaut wings the hard way—by flying the X-15 rocket airplane to altitudes above 50 miles—and with Christopher Columbus Kraft, director of the Johnson Space Center, the aeronautical engineer who directed the flights in Project Mercury. Kraft was a key figure in the bold decision to send *Apollo 8* to lunar orbit in December 1968, a move that may have forestalled a Russian manned circumlunar flight.

At the far edge of the runway, the jumbo jet was warming up with a thunder of its powerful engines. On the orbiter's flight deck, which is palatial compared with the cramped cockpit in Apollo, the test pilots were ready. Fred Haise sat in the left-hand seat as command pilot, Gordon Fullerton in the right-hand seat as pilot. They were talking via radio with Fitzhugh Fulton, the 747 crew commander, below them; the Orbiter Flight Control a thousand miles away at the Johnson Space Center in Texas; and to Ground Control here at Dryden.

Since the early days in Project Gemini, the two-man spacecraft that followed the single-seater Mercury, manned spaceflights have been controlled from Houston, instead of from Florida, where they are launched. This was the result of a pork-barrel decision in Congress to give the southwest a piece of the action in space; it had no technical or logistical justification. One of the prime movers in establishing the astronaut training and flight control center on the Gulf plain below downtown Houston was then-Senator Lyndon Baines Johnson of Texas, whose name the center now bears. The tradition of launching spacecraft from Florida and turning the flight control over to Houston as soon as they leave the pad would be continued in the Shuttle program.

Image Control

Although there was an early effort by NASA publicists to depict the astronaut as a Jack Armstrong, all-American-boy type of hero, constantly being "honed to a fine edge" to perform the daring feat of spaceflight, the image never worked very well to enhance the space program, and it embarrassed the men themselves. The effect of it, in fact, was to alienate this group of intelligent, highly trained, skillful, and strongly motivated men from the taxpayers. The public simply could not identify with such a cardboard hero image at any point, and this led to a lack of identification with the space program itself. As the great adventure in space became routinized, technically intensive, and scientifically abstruse, most of the public lost interest in it.

Some of us who have observed the space age from its beginning in this country have concluded that the effort to glorify the astronauts and present them as supermen embarrassed them and the program. What they really represented was the norm of a cross section of capable, technically oriented and aggressive young men in American society. There were hundreds of thousands of men like them, just as capable, with the same potential. The fact that they represented predominantly Anglo-Saxon, North European genotypes was a reflection of a preferential military selection process that has tended to favor these types. This would change in the Shuttle era.

In the Mercury-Gemini-Apollo era, the predominant image of the astronaut was essentially that of a military test pilot, despite an infusion of young scientists into the astronaut corps. The heroic aspect of the astronaut image created in Project Mercury eroded as space missions became more routine.

By the mid-1960s, the press and much of the public perceived the astronaut as a test pilot approaching or in middle age, with a wife who worried about him, both scrimping to put their children through college. Despite their heroic aura and the risks of spaceflight, these men earned less than commercial airline pilots of comparable flight experience.

It was not until some of them left the corps of astronauts that the hero image paid off in the form of corporate executive jobs, government posts, or high political office. Several of the Mercury Seven astronauts, however, gained financially in motel and banking enterprise and in real estate development.

In the Shuttle era, crews would represent a broader spectrum of the American people. The selection process characterized by racial and sex preference inevitably would break down in response to the social changes of the 1970s.

Initially, however, flight testing of the Shuttle would be done in the test pilot tradition of Mercury, Gemini, and Apollo.

The crews of the *Enterprise* were a case in point. Fred Wallace Haise, Jr., 43 years old, was one of 19 astronauts selected in 1966. He was a trim, brown-haired test pilot of medium height, a native of Biloxi, Mississippi. He and his wife, Mary, also from Biloxi, had four children, the oldest a 21-year-old daughter.

There were four terrible days in April 1970 for Mary Haise and the children after the oxygen tank blew up in the service module of *Apollo 13* en route to the moon. Haise was lunar module pilot on that mission, with James A. Lovell, commander, and John L. Swigert, command module pilot. He had been slated to explore the Fra Mauro formation with Lovell, but the landing had to be aborted.

To bring them back alive, Houston ordered the crew to adjust the flight path of *Apollo 13* so that it would swing around the Moon and come straight home. The horror was the dwindling oxygen supply. Only the use of the lunar module as a lifeboat saved the lives of the crew.

Haise was graduated in aeronautical engineering from the University of Oklahoma and had an honorary doctorate in science from Western Michigan University. He was a member of the Society of Experimental Test Pilots and a Fellow of the American Astronomical Society. He had won the Presidential Medal of Freedom and NASA's Distinguished Service Medal. He was the outstanding graduate in his class at the Aerospace Research Pilot School at Edwards. Since 1952, when he entered Naval Aviation Cadet Training at Pensacola, Florida, he had accumulated 7,600 hours of flying time.

In the seat beside Haise was Charles Gordon Fullerton, age 40, also a graduate of the Aerospace Research Pilot School. Fullerton was a lieutenant colonel in the Air Force. He had been selected as an astronaut in the Air Force Manned Orbital Laboratory (MOL) program, which was abandoned in 1969. He then transferred to NASA. Fullerton had more than 7,000 hours of flying time.

A native of Rochester, New York, Fullerton was about six feet tall, slender, and balding, with a quick, incisive manner and the gift of coming to the point in a report. He and his wife, Marie, had a son and daughter. Articulate and analytical, Fullerton had a master's degree in mechanical engineering from the California Institute of Technology, where he did his undergraduate work. He had gone on active duty with the Air Force in 1958 after working as a design engineer for the Hughes Aircraft Company at Culver City, California. He held Commendation and Meritorious Service Medals from the Air Force.

A native of Abilene, Kansas, Joe Henry Engle, 45, was a graudate of the Experimental Test Pilot and Aerospace Research Pilot Schools of the Air Force. He was one of the 19 astronauts selected by NASA in 1966. One of the most experienced pilots in the world, Engle had flown 130 different kinds of aircraft. He and his wife, Mary Catherine, had a son and a daughter.

Richard S. Truly, 40, was a Navy commander. He was

The two crews for Shuttle approach-and-landing glide tests at Edwards Air Force Base/Dryden Flight Research Center, California, in 1977 were (left to right) C. Gordon Fullerton and Fred W. Haise, Jr., pilot and commander of the first crew, and Joe H. Engle and Richard H. Truly, commander and pilot of the second crew, shown here with Orbiter 101, the *Enterprise,* at Rockwell International Space Division's orbiter assembly facility at Palmdale, California.   (NASA)

born in Fayette, Mississippi, received a bachelor's degree in aeronautical engineering from the Georgia Institute of Technology in 1958, and entered flight training through Navy ROTC. He too was a graduate of the Research Pilot School and had remained at Edwards as an instructor until selected as an astronaut for the MOL. When that Air Force project was washed out as duplicative of NASA's program, Truly transferred to the NASA corps of astronauts. He and his wife, Colleen, had two sons and a daughter.

The 747 was now rolling down the runway, gathering speed. It roared by us with a rush of wind and a pall of acrid exhaust. The *Enterprise* was perched on its back like a sparrow on a hawk. Effortlessly, this strange lash-up lifted into the air, dwindled, vanished. There was a brief ripple of applause, and the attention of the crowd dispersed; little groups formed, and children ran around in games of tag. Loudspeakers mounted on poles relayed a semiaudible account of the orbiter's preflight checks as the jumbo jet climbed beyond the mountains and became momentarily visible again as a flash in the sun, with a faint white contrail. Testing of the orbiter's aerodynamic surfaces and its numerous subsystems would continue for 45 minutes before the glide test started.

The First Moment of Truth

The approach-and-landing tests would tell whether the Shuttle-Orbiter would fly as a glider. Wind tunnel tests predicted it would. But it had to be proved. The *Enterprise,* at 150,000 pounds, was the heaviest glider in the world. It had evolved that way under the twin pressures of reducing costs and weight.

These tests were the only part of the flight profile that could be verified before launching the space airplane into orbit. Its full performance at launch as it was boosted up through the atmosphere, its behavior in orbit as a spacecraft, its ability to withstand the heat and buffeting of reentry, and its stability as a heavy glider at hypersonic and supersonic velocities could not be tested short of a full orbital test flight. Only the orbiter's performance as low-altitude glider from 25,000 feet or less to touchdown could be tested in advance. It was the first moment of truth in the Shuttle program.

There was something else that was new about this vehicle. It would be the first spacecraft to make its maiden voyage into orbit with a live crew. The Mercury, Gemini, and Apollo capsules had flown their initial space tests by ground control without a crew. Mercury's first "crewman" was an astonished and indignant chimpanzee named Ham.

At the outset of the Shuttle development program, James C. Fletcher, who had been appointed NASA administrator by President Nixon in 1971, had told the press that in its first orbital test the Shuttle would not be manned. Again, under pressure of holding down costs and moving development along, this policy changed. Mercury, Gemini, and Apollo were ballistic reentry vehicles. Once their retrorockets fired, they were committed to splashdown at the end of a descending path over which the crew had little control. By rolling Gemini in a "heads up" or "heads down" attitude, the pilot could control

the point of impact to some extent—in theory. But attempts to do it were not entirely satisfactory.

Now, the orbiter's landing was another ballgame. Once the craft entered the atmosphere, it had to be flown, and because it had no propulsion during its long descent, the piloting had to be precise. Although it was flown through a highly sophisticated data processing system, the orbiter required a pilot to manage the descent and landing. Flying the orbiter without a pilot would have required extensive alterations in the flight computer software so that the autopilot would respond to radio beacons along the route as a pilot would.

NASA's decision to fly the first orbital mission with a crew reflected the confidence of the agency's design engineers that the orbiter would be made to work the first time—but only if there were a man in the loop. Flying the big vehicle unmanned presented a technical risk the agency wanted to avoid, given the financial pressures under which it labored.

In the approach-and-landing test series, the low-altitude glide testing was being done in the aircraft tradition, with a crew on board. It was a tradition established by the Wright brothers, and one that may have influenced the orbital test decision, for despite its rocket engines, the orbiter was perceived by its makers as an aircraft that could function in space, rather than as a spacecraft that would fly in the atmosphere. This distinction is important, because it contributed to a near-fatal error in judgment in the design of the heat shield.

The orbiter looked like an aircraft. Its commercial aircraft size made it appear enormous for a spacecraft. Its upper and lower flight decks contained ten times as much room as the conical Apollo command module. The orbiter could carry a crew of seven—ten in an emergency. The Apollo CM was a tight fit for three.

Moreover, the aircraft-style fuselage of the orbiter consisted of a cargo bay 60 feet long and 15 feet wide. It was designed to lift 65,000 pounds of payload into low Earth orbit when the Shuttle was launched eastward over the Atlantic Ocean from Florida. The eastward launch took advantage of the impetus of the Earth's rotation, which added about 1,000 miles an hour to the orbiter's velocity when it left the pad.

When the *Enterprise* was first hauled by truck trailer from the Rockwell International assembly plant at Palmdale, California, to the Dryden test center on January 31, 1977, the hundreds of spectators could have sworn that what they saw surely was an airplane, a space airplane.

Captive Flight

The test series began with taxi tests to make certain that the orbiter would remain stable when bolted onto the top of the Boeing 747 jumbo jet from liftoff to touchdown. These tests did not require a crew, nor did the ensuing five "captive" flight tests, during which the *Enterprise* remained attached to the 747. During these tests, the orbiter's rear fuselage was streamlined by the addition of an aluminum tail cone, which reduced wind buffeting on the tail of the 747. The tail cone weighed 2.5 tons, simulating the weight of the three rocket engines that would power the orbiter at launch.

The five unmanned captive tests were followed by three more, two crewed by Haise and Fullerton and the third by Engle and Truly. On these, the crews had to be content with dry-run testing of the aerodynamic controls: the elevons (ailerons) on the wings, the big rudder and speed brakes (segments of the rudder that open forward to increase air-resistance braking), and the body flap, a rectangular spoiler that is lowered from the underside of the aft fuselage to increase air resistance.

Although the pilot can cause the aerodynamic surfaces to be moved by moving his hand controller, he does so essentially as a programmer. He uses the hand

Orbiter 101, the *Enterprise*, is hoisted atop the Boeing 747 carrier aircraft at NASA's Dryden Flight Research Center, Edwards Air Force Base, California, preparatory to a three-hour test flight November 15, 1977.   (NASA)

Enterprise was carried aloft for the glide tests by a modified Boeing 747 transport that NASA bought second hand from American Airlines. The orbiter is shown perched piggyback on the 747. A tail cone has been installed on the space plane's aft engine section to smooth the air flow and reduce aerodynamic buffeting. (NASA)

controller to instruct the flight computer through which the orbiter is controlled. It is tempting to think of the computer system as a brain, but that is exaggerating its function, which is more akin to that of a nervous system. The pilot is still the brain, and the human brain always is in charge. The data processing (computer) system and the black boxes through which it actuates the aerodynamic surfaces and gas thrusters to steer the orbiter function in a way that is analogous to that of the human central nervous system but is immeasurably faster. The computers respond to sensing devices which monitor the ship's course in orbit or during ascent and descent—the course that has been programmed into the computer. The computer instantly reacts to any flight aberration by flashing corrective steering commands electronically to the reaction control system (RCS) jets; to the elevons on the wing; to the rudder, speed brakes, and body flap; and to the main engines during ascent to orbit.

The "muscle" that moves the massive aerodynamic surfaces is a hydraulic power system, like that of a conventional jet aircraft. However, because the orbiter is a glider in the atmosphere (except during powered as-

cent), it does not carry jet engines. Hydraulic power is provided by three auxiliary power units (APUs). These are turbine engines in the aft fuselage which run pumps. They are used during ascent, turned off during orbital flight, and turned on again for the gliding descent.

The data processing system consists of four computers which work together, plus a fifth which remains passive as a backup. If one of the four fails, the other three can turn it off; if all four fail, the fifth takes over and flies the ship. If the fifth should fail, the orbiter would crash. It cannot be flown without at least one computer operating. Although the pilot can fly the orbiter manually, the pilot's hand controller has no mechanical linkage to the steering devices. It simply signals the computer, which then issues commands through the data processing system.

Free Flight

The first free flight was made on August 12, 1977, by Haise and Fullerton with the tail cone on to reduce wind buffeting. Some 70,000 spectators swarmed over the hills and desert floor to watch. Traffic was backed up for miles on Roseman and Lancaster Boulevards leading to the test site, the seven-mile runway on Rogers Dry Lake.

The 747 took the *Enterprise* up to 22,000 feet above the desert floor. At 8:48 A.M., Haise pressed a button that detonated explosive bolts. The *Enterprise* soared free for the first time. It went up and to the right while the 747 turned left in a shallow dive. At 8:53 plus 51 seconds, the *Enterprise* touched down on the lake-bed runway. The flight was pronounced "beautiful." Not merely successful, but "beautiful." As a glider, the *Enterprise* was magnificent.

Two more flights with the tail cone were flown on September 13, with Joe Engle and Dick Truly as crew, and on September 23, with Haise and Fullerton. On October 12, Engle and Truly made the first free glide with the tail cone off. They came down more steeply and faster than they had with the cone on, but still the orbiter handled well, they said.

Although these first four glide tests looked good to untrained observers, all had actually overshot the touchdown point. Haise and Fullerton had overshot by three-quarters of a mile; Engle and Truly by 680 feet on the second flight; Haise and Fullerton by 786 feet on the third; and Engle and Truly by 1,000 feet on the fourth. The overshoots were not critical on the lake-bed runway, nor would they be considered so on the three-mile concrete runways at Edwards or the Kennedy Space Center. Still, in an emergency, pinpoint accuracy might make the difference between a safe landing and a crash. It remained to be seen whether the *Enterprise* were capable of it.

The Fifth Free Flight

Pinpoint accuracy of touchdown became the objective of free flight number five, which was scheduled for October 26, 1977. Haise and Fullerton were assigned to fly a straight-in approach of 11.5 miles after separating from the 747 at 17,000 feet. They would fly without the tail cone and with dummy rocket engines in the tail.

For this test, *Enterprise* would be landed on an asphalt-concrete strip—the 22-04 runway at the Dryden test center. It was as long and as wide as the Shuttle runway at the Kennedy Space Center.

The crew was instructed to glide at 22° on their initial descent path. This is a very steep glide slope, a dive bomber slope, but it was the slope that operational Shuttle flights would follow on their approaches to Kennedy Space Center.

Coming down at 22° would increase the velocity of *Enterprise* from 245 knots at separation from the jumbo jet to 290 knots. At 11,700 feet, Haise would open the speed brakes to slow down. At 2,000 feet, he would close the brakes and flare the vessel to flatten the glide to a slope of 3°, the normal slope for an aircraft on its final approach.

At 200 feet, he would flare the vessel again, to reduce velocity and adjust the rate at which the vessel was sinking so that the wheels would touch the runway at the touchdown point. The glide slope would be flattened out to 1.5°. The orbiter would look as though it were floating.

At this stage, Haise would drop the landing gear and touch down at 185 knots, applying the wheel brakes hard to halt the orbiter at the 10,000-foot marker, with a mile of runway to spare.

So read the preflight scenario. Would the flight follow the script?

By 8 A.M., the morning of October 26 had turned sunny and mild. The desert Tea Party was in full swing. Prince Charles was standing and chatting at ease, and as long as he stood, everyone else stood, too. Everyone seemed to be holding a tea cup.

Dryden Ground Control called the Boeing and the orbiter: "905 and *Enterprise*. It's Go for takeoff." A thousand people edged up to the clothesline barrier along the runway and peered southwest to catch a glimpse of the spaceship and its carrier, NASA aircraft 905.

The 747 and its 75-ton papoose came rolling toward us, gathering momentum with the incredible quickness of jets. The gross weight of the joined vehicles was announced as one-half million pounds. The big jet lifted it easily at 54 seconds past the hour.

Without the tail cone, the tail surfaces of the 747 were heavily buffeted by the orbiter's air stream.

On this glide test, *Enterprise* is carried aloft without the streamlining tail cone. The main engine nozzles visible aft are mock-ups designed to simulate the aerodynamic effect of the engine configuration during glide.   (NASA)

"How do you like the ride up there?" Fitz Fulton, in the jet, called to Haise and Fullerton.

"It's a little tough to drink the coffee," Fullerton replied. "We're getting, as reported, a lot of this left and right stuff, Fitz. A bit irregular, but constant."

"Passing through 10,000 feet," announced the public address loudspeaker near the Tea Party compound. "Haise and Fullerton will attempt to grease the orbiter's main landing gear on Edwards' main runway at the 5,000-foot mark and plan to have the bird stopped by the 10,000-foot mark. Such a precision landing is a forerunner to the Shuttle-Orbiter's operational landings on KSC's 15,000-foot runway."

The announcement rang with confidence. As the "world's biggest biplane" reached 16,000 feet, Houston called for a 15° left bank to a heading of 340°. Haise ran through the flight control tests, and Houston watched the telemetry results. "Everything looks good to us here," Houston said. Mission Control then called for a turn to a heading of 042 to align the orbiter's descent path with the runway. Pushover, the start of descent, was scheduled at 19,300 feet, and the joined vehicles would then begin to accelerate downward in anticipation of separation.

"Up and over the hill here," Haise remarked. "I'm glad the approach isn't into LAX (Los Angeles International Airport) this morning." A dense fog covered the Los Angeles metropolitan area, which he could see to the southwest.

Houston: I understand it was clobbered there.
Haise: It still looks like it. We're way down south over the hills now.

The 747 and orbiter were no longer visible; the crowd along the runway waited as though it had a lot of experience waiting. That is mostly what you do when you observe any space activity. The longest wait I remember was for John Glenn to make the first American orbital flight in *Mercury-Atlas 6*, 15 years earlier. We waited two months while everything went wrong except John. Even so, they feared he was a goner when false telemetry made it appear that the Mercury heat shield had come loose on the second orbit.

"Go for pushover," Houston called. At that moment, Haise reported that Auxiliary Power Unit 3 had failed. Inasmuch as power for the flight controls was being supplied by APU 1, and APU 2 was still functional, there was no emergency. But on an actual space mission, such a breakdown would have required Houston to bring the orbiter back early.

"Go for sep [separation]," Houston called. Haise detonated the attachment bolts. With a lurch, the *Enterprise* sailed away from the 747. The orbiter dived steeply, flared and seemed to float, and then settled into its steep descent again. It came suddenly into view of the crowd, looking as though it were diving right on top of us. Haise, at the controls, flared the ship and levelled off. The left and right dual wheels of the main landing gear came down. The two nose wheels came down.

The *Enterprise* was suddenly rushing by us with a whoosh of wind. No jet engine roar. Just a rush of air. The main gear tires touched the runway, shrieking; immediately, the orbiter was back in the air again. That was not according to the script.

As the ship rushed by the Tea Party, it was oscillating left and right, emitting a shrill whistle. But before anyone could realize that something was askew, the ship stabilized and sank once more to the runway, only to bounce off the concrete a second time. It rose slightly into the air and came down a third time, rolling fast, nose up. It just didn't seem to want to land, and I could imagine Haise trying desperately to force it to stay on the ground. He was fighting a balloon of air beneath the big delta wing that no one had bargained for.

For eight long seconds, the *Enterprise* rolled nose up. Finally, the nose wheels touched. The crowd cheered and applauded. Far beyond the main body of observers, the ship halted, rather suddenly. It rested like a fat bird, huffing and chuffing with escaping gases from its power systems.

The descent had taken 2 minutes and 1 second, about 6 seconds longer than calculated before the flight. The orbiter had rolled for 50 seconds after its nose wheel came down. It was the roughest landing of the free flight series. Although it was slated as the last test, NASA officials had to consider the advisability of repeating it, for this was a far cry from the pinpoint landing they advertised.

The postflight news conference was held in a small auditorium at the Dryden Center, adjacent to a cafeteria, where several busloads of schoolchildren were having lunch, running about, screaming, giggling, wrestling, crying, and getting lost. As part of its effort to integrate itself with the rest of the country, NASA has opened up its space centers to public visiting. Only key areas are off limits. The scene at Dryden was a far cry from the fanatic secrecy of the late 1950s, when newsmen had to peer through spy glasses to find out what was going on at Cape Canaveral and the agency regarded the press as its adversary and the public as a nuisance.

Fred Haise confessed to assembled media reporters that the landing was "a personal disappointment." The *Enterprise* had come in too fast. When it reached the runway threshold, its velocity, which should have been no more than 280 knots, was over 290.

"I was sure the speed was going to start cranking off," Haise said. "And it didn't. The speed was not dropping off as we had seen in simulation."

The first bounce caught the pilots by surprise. "All at once I found myself back in the air again," Haise said. "And this time it skipped. And when it skipped, whether it was a rolling input or I induced it, I'm not sure which, I then got into what we call a PIO—pilot-induced oscillation—where I was chasing the bank angle laterally for about three or four cycles before Gordo called to me and told me to relax, which is the normal thing you got to do to stop that kind of thing. So I let off the stick and it stopped."

Fullerton added: "I believe our gliding performance is better than the engineers predicted, and that's what made us fast when we got up to our planned touchdown point, and that led to the skip and the oscillation and so forth."

What did Prince Charles have to say? the crew was asked.

"He is a pilot," said Fullerton, "and [at Houston] he

adapted very rapidly to the somewhat unconventional control stick that we have in the orbiter and made a couple of very acceptable landings. He's a very personable man to speak to, and we talked about flying most of the time."

Haise added, "Actually, even though I wasn't too happy, I think Prince Charles was. Because he had encountered this sort of thing on his second approach in the OAS [orbiter atmospheric flight simulator]. He had gotten into a little lateral oscillation and he was pleased to see that I duplicated that."

Test Manager Deke Slayton commented that although the landing was not optimal, it was no worse than many made by commercial airline pilots. Slayton was satisfied with the orbiter's performance. As far as he was concerned, the test series was ended.

That was not the view at Houston, however. Films and simulations of the landing were run for the next two days to make sure there was nothing in the orbiter's flight performance that engineers and pilots didn't understand. Despite the overshoot and double bounce, Fullerton had pointed out, "we still had 2,000 feet of runway to spare." So there would have been no hazard if the landing site had been the Kennedy spaceport.

"This is really a very tiny first step when you consider the entire flight envelope of the orbiter," Fullerton said. "It's a very small corner. We have a long way to go, a lot of problems to solve."

Houston confirmed Slayton's view that the orbiter's glide behavior on approach and landing was sufficiently well understood that no more glide tests were needed. That part of the flight envelope had been explored.

The next great moment of truth would come when a Shuttle-Orbiter was launched on its first manned orbital flight from the Kennedy Space Center in Florida. At the end of 1977, no one in the space agency could predict with any certainty when that would happen. It had been scheduled in 1972 for the end of March 1978, following the horizontal flight tests of 1977. By that time, however, the production of the two development orbiters had fallen a year behind schedule. Orbiter 101, the *Enterprise*, had become too heavy to perform adequately as an orbital vehicle, although its performance in the lower atmosphere as a glider had confirmed the glide expectations of the designers.

Orbiter 102, named *Columbia,* was being manufactured at Palmdale as the first Shuttle flight article. No one knew when it would be ready to fly, but NASA headquarters selected a tentative target date of the end of March 1979.

As it turned out, the ten months of approach-and-landing tests were only the beginning.

Orphan of the Storm

The approach-and-landing tests confirmed some of the confidence that NASA had exhibited in selling the Shuttle to Congress and the administration in 1972, but the series, for all of its photogenic opportunities to display the new space plane, had proved only one thing. NASA and its prime contractor, Rockwell International, had succeeded in building the world's heaviest glider.

While this achievement was hailed as a triumph of American know-how, dark clouds were clustering about the Shuttle program. Behind the scenes, the notion that this new invention could be built almost entirely within the state of the art, largely from "off-the-shelf" parts (which either existed or could be readily fabricated) was eroding away.

True, the top echelon of NASA officialdom had qualified their presentations to Congress with caveats about the engines, the heat shield, and the elaborately computerized flight control system the orbiter would require. These aspects of the vehicle involved technological ad-

vances. To that extent, they exceeded the state of the art. But nowhere in the positive attitudes that Shuttle directors displayed before congressional committees was there a suggestion that these difficulties, if encountered, could not be overcome by the ingenuity that had produced Apollo. Nowhere did anyone say, "We have tough sledding ahead; it may hold us up a year or two and raise our costs."

From 1972 on, Shuttle briefings for the press rang with optimism. The theme was: We have the technology; we have the skills; we know how to build it; it will reduce space transportation costs to a reasonable level. The space agency's confidence seemed to be derived from its projection of the Shuttle as simply a space-adapted airplane. You install rocket engines in the tail and paste a heat shield on it, and lo—you have a space airplane. But . . . do you?

Building a ballistic reentry vehicle like Apollo was a problem of a lower order of magnitude. Modifying an airplane to function as a spacecraft in one medium and as an airplane in another was like devising a submarine that could also be flown as a dirigible. It was possible in theory if the problems of transition from one medium to the other could be resolved. It was this problem—transition—that was to delay Shuttle development three years and increase its cost by 60 percent.

During 1972 and for five years thereafter, the real difficulty of creating the Shuttle was masked by the Simple Simon conception of it as a space-adapted airplane. This idea led NASA and its contractors to devise an unworkable heat shield whose integrity remained in question until virtually the eve of the first manned orbital test flight. It was not until after the glide tests in 1977 that the real nature of the Shuttle became publicly apparent. It was a winged spacecraft: a new invention the like of which, despite some ancestral similarities, had never existed before.

The lack of a clear definition early on of what the Shuttle was or was going to be and a clear assessment of the order of difficulty that would be encountered in building it accounts for the unrealistic development cost estimate that NASA gave Congress. In my view, this also explains the agency's failure to test the new high-pressure hydrogen-oxygen engine components, especially the fuel and oxidizer pumping systems, separately in the conventional way before bringing the fully assembled engine to the test stand, as well as its failure to test the new heat shielding for strength in ballistic reentry tests, as early missile nose cones were tested.

The Shuttle development program required the construction of two orbiters: 101, the *Enterprise,* and 102, the *Columbia.* Modifications of these development orbiters found necessary by flight testing would then be incorporated in later production orbiters.

The fact soon became apparent that NASA could not build an operational Shuttle for $5.15 billion as advertised. The agency not only underestimated the technical problems but overestimated the expertise required to solve them without the long, agonizing process of test failures, breakdowns, parts shortages, and cost overruns that have plagued every rocket development program since the German V-2 of World War II.

Moreover, the Shuttle posed a new flight experience, which could be anticipated only by simulation on the ground and with limited data from the X-15 and related test vehicles. With the glide tests of 1977, NASA reached the end of the known; from then on it entered the unknown, where the risk was high. Still, the logic of reusable spacecraft was persuasive; it would not only cut costs of putting satellites in orbit but also put Americans back there to match the continuing Russian presence.

By minimizing, or understating, or failing to concede, the order of difficulty, the NASA Shuttle directorate gave the project a rather low profile. In 1972, *Apollos 16* and *17* were visiting the Moon, and although the scientific chores of the astronauts on the lunar surface had become so mechanical and abstruse that much of the public lost interest in them, these missions still retained some of the high drama of earlier landings. Also in 1972, Sky-

Early orbiter design evolved largely from these famous lifting bodies built for testing by NASA and the Air Force. Left to right are the X-24, the M-2, and the HL-10, all considered orbiter prototypes.   (NASA)

lab, the space station as large as a five-room house, was being completed for launch to test the proposition that a large, comfortable, well-equipped station in orbit was good for the country: it was good for science, good for the investigation of manufacturing opportunities in space, and good politically because it was five times bigger than anything the Russians had.

As it evolved between the lunar Apollos and Skylab, the Shuttle seemed far down the road indeed. Vaguely, it was linked to the servicing of a permanent or semipermanent space station, but that was uncertain. At any rate, during the first five years of its development, the Shuttle commanded hardly any more attention among the general public than an interstate highway project.

The Shuttle? people would ask. What's that? The public information apparatus of NASA had its hands full trying to explain it. This difficulty was exacerbated simply because the mass media of communication, radio, television, and newspapers, were hardly interested. It was not until the *Enterprise* began its free flights in the Mojave Desert that the television networks dispatched their famous folk to the scene to tell the people what the Shuttle was. By that time, it had undergone quite a metamorphosis. In fact, the *Enterprise* no more resembled the original concept of the Shuttle circa 1970 than the QE II resembles Noah's Ark. There had been a great deal of evolution since then, involving a series of deleterious mutations caused by the Office of Management and Budget (OMB), the executive budget agency of the United States.

Since I began to cover space programs on a regular basis, in 1961, I have been constantly impressed by the fact that NASA's program is mainly a response to a national mood; it does not generate that mood. Mercury, Gemini, and Apollo were quasi-military responses to the initial Russian lead in space technology, which had implied a military threat to the security of the United States and at the same time suggested the existence of a latent technical superiority that was coming to the fore in the Soviet Union. These concerns were the mainsprings of the hysteria that gripped the nation in the early 1960s, created the semifictitious flap about the missile gap, and persuaded Congress to write a blank check to finance

John F. Kennedy's call for a manned lunar landing within the decade.

With Apollo, NASA was reacting to the space race, an exercise in the chauvinistic competition of derring-do that had not been equalled since the heroic period of antarctic exploration in the first decade of the century, when Scott lost the race to the south pole to Amundsen and perished.

Initially, the Shuttle was an outgrowth of the lunar landing euphoria; but before the project really took shape, the mood died away, reversed itself, and was supplanted by a profound indifference.

Apollo, an ad hoc program to reach the Moon, was not viable as a long-term transportation system. It was too costly. Once the lunar landings were accomplished Apollo's usefulness ended, although its flight hardware continued to be used in the Skylab experimental space station and the 1975 Apollo-Soyuz test project.

With the end of Apollo in 1975, a principal, motive for the creation of NASA—manned spaceflight—could be realized only with the Shuttle. Manned flight required an elaborate infrastructure: NASA's worldwide system of tracking and relay stations and its launch center in Florida, development center in north Alabama, and mission control center in Texas. The Shuttle became the reason for being of this billion-dollar array and the corps of astronauts. It was the only vehicle NASA could call its own after the Apollo-Saturn era ended in 1975. The launchers it used—the Atlas, Delta, and Titan—were derived from military missiles. Without the Shuttle, there was no manned flight program; without a manned flight program, NASA's function would be so severely curtailed that its survival as an independent agency would be doubtful.

It was not difficult, therefore, to understand why the agency was selling the Shuttle as hard as it could and why it so readily compromised the vehicle's reusability feature by degrading the original concept to that of a rocket-boosted glider. This was done, as I will relate later, to conform to the funding policy of the Office of Management and Budget as a means of survival.

The executive budget agency didn't know how to build a Shuttle, but it had the power to control the funding of the project and thus to dictate the construction strategy.

In the days of Project Mercury, NASA had an administrator by the name of James E. Webb. He was not a scientist or an engineer. He was a promoter with federal budget experience and the gift of making manned spaceflight appear to be the greatest event since the creation. After Webb retired, his genius was lost to NASA and never recaptured. The agency slid into the gravitational well of federal bureaucracy—the black hole from which no aspect of creativity, even imagination, ever escapes.

Now the agency had the Shuttle, but it had no Jim Webb to sell it as he had sold the Moon. Instead, it went out and hired financial consultants to prove that it was cheaper to put up satellites with a reusable Shuttle than with throwaway rockets. Who cared? The benefit to the public was remote and intangible. This was corporate selling. It sought to justify a new development by its cost-effectiveness, like a supermarket.

The economic rationale for the Shuttle was not the main issue and disputable at best. The real reasons for building the Shuttle were political and military, the same as the reasons for building Apollo. As long as the Russians were keeping their men in orbit in Soyuz and Salyut vessels, they were looking down our throats and incidentally conducting space manufacturing experiments

Artist's drawing of satellite deployed by Shuttle-Orbiter.

(NASA)

that might some day pay off in a big way. There was mounting concern that we match that military reconnaissance presence and that potential industrial capability.

The Blueprint

Public interest in manned space exploration began to fade after the first two landings on the Moon confirmed the general notion that it was a lifeless desert. Technically, the lunar voyages were exciting; scientifically, they were abstruse. The astronauts themselves were reduced to laboratory technicians and rock collectors. Their activities were rigidly constrained by an electronic leash that bound them to Mission Control at Houston. There was no adventure here. There were scientific experiments the nature and purposes of which held little fascination for the lay audience.

Although each of the voyages of Apollo contributed some information about the origin and evolution of the Moon and about its connection with Earth, there was no continuity or sense of discovery. True, some of the data took months and years to resolve, yet the general nature of the Moon as a partially evolved planet became apparent quite early. But the fact that here was a vast storehouse of raw materials mankind could use—iron, titanium, aluminum, and thorium—became obscured by the emphasis of the scientific investigators on cosmogonic questions: the origin and evolution of this body. In the welter of debate and speculation that characterized the lunar science conferences during this period, the potential of the Moon was largely ignored. It was as though the principal interest of Columbus' sponsors had been focused on a soil analysis of Hispaniola.

Public interest in the lunar flights faded virtually in direct proportion to the increasing complexity of the scientific experiments. With admirable detachment from an economic point of view, the space agency projected the view of the scientific community, which regarded the Moon as a scientific preserve. It was a view that made the Moon easy to abandon as a practical matter. Its potential wealth and utility for a spacefaring nation was not

generally realized, even as late as the end of 1972, when it was abandoned by the United States after _Apollo 17_ came home.

As a goal in an international chariot race between the United States and the Soviet Union, whatever incentive the Moon might have contributed as a steppingstone to the exploration of the Solar System by manned expeditions was lost. Once you have scored, why go farther? Next stop is the locker room. And that is where NASA headed in 1972 to plan its next ballgame.

However, during that year of miracles, 1969, when we made three trips to the Moon in _Apollos 10, 11,_ and _12,_ the intensity of public interest seemed to demand a program beyond Apollo. A post-Apollo plan was devised for the balance of the century. It projected an interplanetary transportation system that could reach Mars. The initial unit of such a system was a ground-to-orbit Shuttle. Thus, the Shuttle was conceived of as part of a transportation complex that was to extend the range of manned operations far beyond the Moon.

This scheme was already drawn when Neil Armstrong jumped to the surface of the Moon from the last rung of the lunar module ladder and proclaimed: "That's one small step for man, one giant leap for mankind." He had intended to say "one small step for a man," but somehow the article was missing. Although NASA hastily amended the statement, Armstrong was right the first time. In the context of man's opportunity in space, the landing on the Moon was one small step.

It was one small step out of Tsiolkovsky's cradle. After that first small step, it seemed reasonable to believe that man would take longer strides. Still, it was a hard act to follow.

En route across the Pacific Ocean to a conference in the Philippines, President Richard Nixon, accompanied by his national security adviser, Henry Kissinger, paused to welcome Armstrong and Edwin E. (Buzz) Aldrin back from the Moon. According to Kissinger, Nixon was deeply moved by the event and declaimed that we had witnessed the greatest week in the history of the world since the creation.[1]

What, indeed, could surpass the first landing on the Moon? The answer was obvious to the framers of the post-Apollo plan: a manned landing on Mars. But that could not be done with the Apollo-Saturn transportation system. Mars would require a larger, more powerful one. There was one on the drawing boards at NASA's Langley Research Center at Hampton, Virginia.

During his first six months in office, Nixon appointed a Space Task Group to design a post-Apollo program. The members were the vice president, Spiro T. Agnew, who by virtue of his office was chairman of the National Aeronautics and Space Council, an advisory group; Robert C. Seamans, secretary of the Air Force; Thomas O. Paine, the holdover NASA administrator from the Johnson administration; and Lee A. DuBridge, the president's science adviser.

These men were no amateur strategists, and it is clear that their view of the Apollo program was that it was only a beginning. The Apollo-Saturn system was so costly that only quasi-military urgency could justify it. Each lunar mission cost more than $1 billion. It cost more than $300 million simply to launch the Apollo spacecraft on the Saturn 5 rocket. Only the richest society in the world could afford such a system, and then not for long. If manned spaceflight was ever to become economically

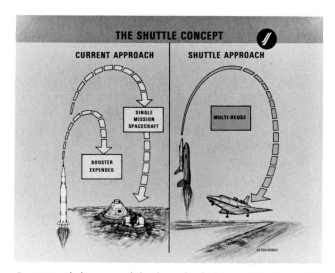

Conceptual diagram of the launch, flight, and landing of the reusable Shuttle (right) compared to the launch of the expendable Saturn 5 rocket and the flight and landing of the nonreusable Apollo spacecraft. (NASA)

rational, it had to be redesigned with reusable vehicles; vehicles that could be used over and over again. Each Apollo-Saturn system could be used only once. The Saturn 5 first and second stages fell into the ocean; the third stage fell on the Moon. The lunar module remained there, along with all the landing paraphernalia, including the $1 million Lunar Rover, the electric jeep the astronauts used on the last three missions. The Apollo command and service modules returned to Earth, but the service module was jettisoned before the command module reentered the atmosphere. Finally, the command module, although recovered from the sea, was never used a second time. It became a museum piece.

DuBridge, a chemist, had been president of the California Institute of Technology. Paine was a top executive at General Electric. Seamans had served in NASA as a deputy administrator during the period when Webb was out selling spaceflight to the country as the salvation of America. From the viewpoint of these gentlemen in 1969, the technological distance from the Moon to Mars was no greater than the distance from the Earth to the Moon had been in 1961, when President Kennedy had called for the manned lunar landing. Within a few years, that view would change and the members of this ad hoc committee would go their separate ways. Paine, a Democratic appointee, would be replaced (1971) by a Nixon appointee, James C. Fletcher, an aerospace industry executive who was president of the University of Utah. Fletcher was to preside over the first four years of Shuttle development.

In September 1969, as the glow of the landing at Tranquility Base was fading, the Space Task Group issued a report: "The Post Apollo Program: Directions for the Future." It began, "The Space Task Group believes that manned exploration of the planets is the most challenging and most comprehensive of the many long range goals available to the nation at this time, with the manned exploration of Mars as the next step toward this goal." The Task Group surmised that a manned landing on Mars "appears achievable as early as 1981." However, in order to ease peak funding in the late 1970s for such a program, the Mars landing could be deferred until the middle or late 1980s.

As the Task Group conceived it, the Shuttle, the basic vehicle in the transportation system, was primarily a space-adapted airplane that operated between a ground terminal and a space station. Its major characteristic would be its reusability. It would be a common carrier. The Shuttle would put transportation to low Earth orbit on a routine and economical basis. It would thus replace the expendable rockets—the Delta, Atlas-Centaur, and Titan 3.

But the Shuttle was limited to low orbits, altitudes no higher than 600 miles. It could not deliver communications satellites directly to geostationary orbit, some 22,300 miles above the equator. An upper-stage vehicle, or space tug, would be required for those missions. The Task Group proposed a tug for interorbital transfers of payloads near the Earth. Chemically powered, it would remain in orbit, docked with the Space Station, when not in use. Larger than the tug would be a third vehicle, a more powerful interorbital transfer stage (ITS), powered by nuclear rockets and/or ion drive engines. The ITS would be designed to

Proposed method of deploying satellites with upper stage boosters (circa 1976). Boosters then considered were solid-fuel spinning upper stages (SSUS), later superseded by the Payload Assist Module (PAM) and the Inertial Upper Stage (IUS) rocket.   (NASA)

move cargo between Earth and lunar orbits or to orbit Mars.

Space stations, composed of joined 8- to 12-man modules, were part of the blueprint, orbiting the Earth and the Moon. Hitched to an ITS, a 12-man space station module or two could carry explorers to Mars orbit, from which a team could descend to the Martian surface in a fourth type of vehicle—a Mars lander. This vehicle would be equipped with aerodynamic systems as well as rockets, to ferry working crews and materiel between the planet surface and an orbiting space transport.

Thus, the interplanetary transportation system could be developed in series, starting with the Shuttle and continuing with the tug, the ITS, the multimodular space stations, and the landers. The rate at which the system could be built would be determined by the amount Congress was willing to invest each fiscal year.

In 1969, a Mars landing in the 1980s did not seem imminent. It allowed 16 years to develop the project, while the landing on the Moon had taken only 8 years from the word go. Publicized in a Sunday supplement mode, its impact was zero. It competed for attention with the reality of the lunar landings, and it came at a time when rising social and political disaffection challenged the relevance of space programs to the national welfare. In this context, going to Mars seemed to serve no purpose. The Russians, who had led the way to low Earth orbit, seemed to be stuck there. They had settled into the groove of supplying a crew in a small space station called Salyut and seeking to set endurance records. It resembled an orbital form of flagpole sitting.

The Magnificent Biplane

From the concatenation of vehicles proposed by the Space Task Group, only the Shuttle materialized. Congress dropped the tug. It was too costly a project to undertake at the same time as the Shuttle, especially with the burden of war costs in Southeast Asia. Similarly, the idea of a permanent space station served by the Shuttle was deferred to an indefinite future. Skylab, the Camelot of space stations, was coming on line as a demonstration

project. It was designed to be serviced not by the Shuttle but by Apollo, and its utility was to end with the end of Apollo. Nevertheless, the imminence of Skylab had the effect of putting plans for a permanent station on the shelf, so that the Shuttle became the only proposal of the Space Task Group blueprint to be implemented.

Before leaving NASA, Paine sought Western Europe's participation in developing the tug, but the European Launcher Development Organization (ELDO) was in disarray as a result of failures with the Europa rocket. Paine found European engineers excited by the prospect of building a tug which could be sold or leased to NASA to operate in conjunction with the Shuttle, but the engineers were not supported by their countries' ministers. Technically, if not linguistically, ELDO was something of a tower of Babel. National entities participating in it tended to do things their own way. Central authority was weak and was often challenged. It was not until ELDO was dissolved and reorganized as part of the European Space Agency in 1975 that a strong central management was established for major projects, on a model somewhat resembling the American plan.

Unable to get a commitment from Europe on the tug, Paine sounded out Japan, but the Japanese were not ready. Like Western Europe, they were moving along the rocket development path, but they were moving more cautiously and, it seemed, less ambitiously. No, if there were to be a tug, the United States would have to build it.

Inasmuch as there appeared to be a mandate to continue developing the capability of manned spaceflight after Apollo, NASA went ahead with the preliminary design for its Space Shuttle as though it were part of the great interplanetary transportation system that the Space Task Group had projected. Even before *Apollo 11* was launched July 16, 1969, on its historic journey, NASA issued requests for proposals (RFPs) to the aerospace industry for a shuttle design, based on the Task Group report. The predominant idea of a Shuttle then visualized was a two-stage vehicle consisting of two airplanes, each powered by rocket engines for the boost to orbit and by jet engines to fly in the atmosphere. It was essentially the concept of the space-adapted airplane.

The Space Shuttle of 1970 was a magnificent biplane.

SHUTTLE IS SIMILAR SIZE TO EXISTING SYSTEMS

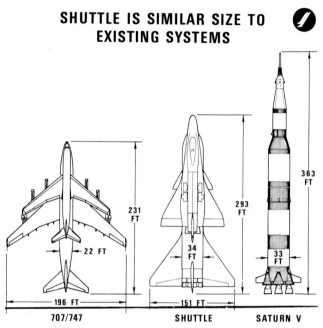

Preliminary designs for the Shuttle (circa 1970) compared with the Boeing 707/747 jetliners and the Saturn 5–Apollo lunar transportation system.   (NASA)

Artist's conception of the two-airplane Shuttle (circa 1970). The orbiter, the size of the Boeing 707, rides piggyback atop a Boeing 747-sized booster. This design was not realized because of high development cost.   (NASA)

The initial designs released by NASA depicted the booster as the size of the Boeing 747 jumbo jet and the orbiter as the size of the Boeing 707. It was big!

The two airplanes would be launched standing on their tails, vertically. One, smaller than the other, would be the vehicle to go into orbit—the orbiter. The larger of the two would be the booster.

At an altitude of 25 or 30 miles, the orbiter would separate from the booster and continue on into orbit under the power of its rocket engines. The booster would then turn around, shut off its rocket engines, and turn on its jet engines to fly back to the launch site, presumably the Kennedy Space Center in Florida adjacent to Cape Canaveral. Both vehicles would carry a pilot and copilot.

There were two aspects of this lash-up that were inescapable. One was that it had to be very large in order to carry fuel. The other was that it had to be reusable, like a commercial carrier. These requirements suggested an initial development cost that would not be less than $10 billion. There was a very serious question whether Con-

gress or the Nixon administration wanted to invest that much in a reusable space machine, especially since the compulsion to demonstrate prowess in space had faded.

Neither the idea of one airplane riding piggyback or back-to-back on another nor the concept of a rocket-propelled aircraft was new in the world.

The rocket plane had been around for half a century but had not evolved beyond experimental design. Nor had it ever been developed to the point where it could fly independently.

In 1928, a German experimenter, Friedrich Stamer, succeeded in propelling a light glider with two small rockets and an elastic rope. A Junkers 33 seaplane was flown with rocket-assisted takeoff at Dessau, Germany, in August 1929.

In September 1929, Fritz von Opel flew a rocket-propelled glider about one-half mile at Frankfurt-am-Main. The German Air Force tested rocket-powered fighter aircraft in 1939–1940. The outcome was the Heinkel 176, but it was replaced by a more reliable and controllable jet engine aircraft.

These experiments are hardly ancestral to the shuttle, but they point out a direction designers were taking. The most imaginative proposal of the World War II period was a rocket-powered intercontinental bomber drafted in 1944 by the Austrian engineer Eugen Sanger and mathematician Irene Bredt. They composed a mathematical model of a 100-metric-ton vehicle 92 feet long, with a 50-foot wingspread. It carried a pilot and a single bomb weighing 660 pounds. Accelerated initially by a rocket sled on tracks, this vehicle would be lifted to an altitude of 162 miles by its rockets and reach a velocity of 3.7 miles a second. That is not enough to put it into orbit, so it would reenter the atmosphere in a shallow dive, skipping out and falling in again like a flat stone skipping over water.

The skips and dips were precisely calculated. During one of the dips, which would occur over a city—New York, for instance—the vehicle would release its 660-pound bomb. If the dips could be matched up with the targets, the scheme might work. The vehicle would continue bouncing in and out of the atmosphere until eventually its velocity was slowed sufficiently to allow it to glide to a landing. The landing site would probably be in Japan, or on some Japanese territory, halfway around the world from the launch site in the Alps or the Danubian plain.

Although this scheme never got off paper, it became widely known as the Sanger-Bredt Antipodal Bomber. By increasing the rocket thrust at launch, Sanger believed, the craft could be flown all the way around the world and landed at the launch site.

The antipodal or global bomber seems to be the prototype of the rocket-boosted glider. In that sense, it is the ancestor of the Shuttle. Along the line of descent is the X-20, the 45-foot delta-wing glider proposed by the Air Force in 1958 as "Dynasoar" (Dynamic Ascent and Soaring Flight). The X-20 was to be boosted into orbit by a Titan 3 rocket. It would reenter the atmosphere by fir-

ing a retrorocket to slow it down and glide to a horizontal landing in the Mojave Desert of California. The project was scrapped by the Department of Defense in 1963, mainly because NASA's Project Mercury, which was flying successfully, made Dynasoar somewhat redundant.

The piggyback idea, one airplane launching another, also had a history. In 1938, the British used a large flying boat, the *Maia,* to lift a smaller seaplane, the *Mercury,* to cruising altitude. The two then separated, and the *Mercury* was able to reach South Africa from the launch site in Scotland without refueling. When World War II started, the duo was operating routinely on *Mercury* flights between Southampton and Alexandria, Egypt. The tactic enabled the smaller seaplane to become airborne without expending fuel, thus extending its range. The "mother ship" *Maia* then returned to the airport or continued on a mission of its own.

France and Russia also experimented with piggyback flight. The U.S. Air Force and the National Advisory Committee on Aeronautics (NACA), which was absorbed by NASA, experimented with smaller craft carried by big ones. The basic idea was staging. The multistage aircraft, curiously enough, evolved before the multistage rocket. The Bell X-1 and X-2 and North American Aviation's X-15 were carried aloft by bombers.

No long step was required to develop a two-stage Shuttle. The basic form of the double airplane was already being considered by NASA in the fall of 1969, when the space agency's Office of Advanced Research and Technology invited contractors in the United States and in Europe to a conference on Shuttle design. The meeting was held at the Smithsonian Museum of Natural History in Washington October 16 and 17, 1969.

The conceptual outline (Phase A) of the project was described by George E. Mueller, NASA Associate Administrator for Manned Space Flight, as "a Shuttle that takes off from Earth, makes rendezvous in orbit [with a space station], and returns to Earth with a minimum of any real control from Earth. It essentially needs only refueling to begin another trip.

"This approval is now before the President of the United States for his final agreement as to whether we ought to proceed to Phase B, the specific, preliminary design."

As defined by Mueller, the goal of Shuttle development was a reduction in space operating costs from $1,000 a pound for payload delivery to orbit by the Saturn 5 rocket to "somewhere between $20 and $50 a pound."

"If we succeed," Mueller said, "we can open up a whole new era of space exploration."[2]

A Welter of Designs

The aerospace firms presented their designs. Most were two airplane stages of different sizes, but all were large. Convair Division of General Dynamics Corporation, San Diego, offered a booster airplane 240 feet long. It would carry aloft an orbiter 174 feet long. The combination was to be capable of hoisting a 25-ton payload to an orbit 270 nautical miles high. Launch propulsion would be provided by hydrogen-oxygen engines in the first stage, which would fly back to the launch site after the orbiter separated from it at an altitude of 31 nautical miles and continued on into orbit under the thrust of its own hydrogen-oxygen engines. The cost of development was estimated at $5.2 to $5.5 billion.

Lockheed Aircraft Corporation also offered a two-stager. Because of internal propellant tanks, the stages were huge in all designs. Lockheed's booster was the size of the Boeing 747 jumbo jet. The orbiter was about the size of the Boeing 707.

Other two-airplane designs were presented by Martin Marietta, Denver Division; McDonnell-Douglas; and the Space Division of North American Rockwell Corporation.

Rockwell's booster was 263 feet long, with a 230-foot wingspan. Its orbiter was 202 feet long, with a 146-foot wingspan. These designs called for a Shuttle weighing close to four million pounds at launch in order to deliver 25 tons of cargo to low Earth orbit.

Prophetically, Lockheed's presentation raised a caveat about the heating problem any orbiter would encounter in reentering the atmosphere. "There are strong uncertainties in estimating the heat loads at hypersonic reentry," warned the Lockheed presenter, Maxwell N. Hunter. Short of a full flight test, he said, "the basic turbulent heating theory . . . cannot be simulated."[3]

Preliminary reentry vehicle design.   (source unidentified)

The warning was prophetic indeed. Not many NASA people would remember it ten years later, when the problem of dealing with reentry heating became the number one problem in completing the first Shuttle flight article, Orbiter 102, *Columbia.*

The uncertainty persisted right down to the launch. It did so because the Shuttle was the first manned space vehicle to bypass the tradition of making its first space test unmanned. Thus, there was no opportunity to examine the reentry heating problem with an unmanned test.

Several European space interests were represented at the conference. A spokesman for the French Centre National d'Études Spatiales, C. Bigot, proposed that a space station be placed in an equatorial orbit to be served by the Shuttle. If that were done, the Shuttle could be launched most economically from the French space launching center at Kourou in French Guiana, 5° north of the equator. It would take less fuel to reach an equatorial orbit from there than from Cape Canaveral or the Kennedy Space Center, at 28.5° north latitude.

Equatorial orbits have definite advantages. At the synchronous altitude (22,300 miles), a satellite orbiting the equator falls around the Earth at the same rate as the Earth's rotation, so that it seems to remain in one place. This is an ideal arrangement for communications satellites. It was obvious to the French, as to everybody else in the space business, that there was more profit in communications satellites than in any other spinoff of space technology. Kourou definitely had a future as a commercial launch site, but NASA was not interested in it in 1969.

French studies also favored a smaller vehicle than the monsters designed by the American companies. One French proposal had the booster airplane take off from a conventional airport runway. At altitude, it would be accelerated by rockets to seven or eight times the speed of sound. Then the second stage would separate, fire its rocket engines, and go into orbit, while the booster turned around and flew back to the launch site. The payload for this design was rather meager, about 6,000 pounds. But that would be sufficient to service a space station, if that were the Shuttle's main function.

NASA's view of the Shuttle was finally spelled out by the director of the Office of Advanced Research and Technology, Adelbert O. Tischler.

"The system of which we speak is really a rocket propelled aircraft," Tischler said. "The last such machine built in this country was the X-15. It was conceived and built in the early 1950s and operated over a range of flight altitudes and speeds which have never been exceeded by any other aircraft."[4]

Tischler said that the X-15 had reached a peak altitude of 67 miles. Only 33 more miles and it might have achieved orbit; yet it was not designed as an orbital vehicle. It was a rocket plane, using the B-52 bomber as its first stage.

In the realm of aerodynamic effects, however, the Shuttle posed a more complex problem than the X-15 or any other aircraft. The booster stage had to be a hypersonic aircraft capable of carrying an enormous cargo faster than any aircraft had flown before. The orbiter stage had to be a reentry body, capable of controlled flight through the entire range of trans-sonic velocities and manageable in the lower atmosphere at subsonic velocity.

Tischler regarded the challenge of the Shuttle as truly remarkable. It was a task, he quipped, comparable to that of putting a ring in the nose of a bull from the wrong end of the bull. "This is going to be a nasty job," he said, "but I'm imbued with the conviction that it can be done if we can work our way through the bull."

The Money Problem

The first obstacle encountered in working through the bull was cost. In the decade between the start of Apollo and the start of the Shuttle, the cost of technological developments had become a critical issue. On the one hand, it was seen as competing with social amelioration programs—taking bread from the poor. On the other, the need for or "relevance" of appropriating billions for a Space Shuttle was fully acceptable only to a technically minded minority. It was unacceptable to many articulate critics who regarded the alleviation of economic distress as the nation's all-engulfing priority. Programs of high technology in general were being challenged by the spread of a militant humanism among young intellectuals and their campus followers, who considered themselves the vanguard of an American cultural revolution. These young militants branded virtually every form of technology, from spaceships to modern agriculture, as antihumanistic.

Beyond these attitudes existed a real fiscal dilemma in the changing priorities of the national budget that would subject any major new development to constraint. The means of financing new efforts in science and technology were becoming curtailed by the rapid growth of federal entitlement programs such as welfare, education, Medicare, and Social Security. By the early 1970s, the slice of the budget pie allotted to NASA and the National Science Foundation had dropped from 2 to 1 percent.

This effect was described by Representative Edward P. Boland (D-Mass.) at the 1978 annual meeting of the American Association for the Advancement of Science.

Boland was chairman of the House Independent Agencies Subcommittee. In 1968, he said, only 50 percent of the federal budget was committed to the entitlement programs. But by the 1978 fiscal year, the commitment had risen to 75 percent.

Not only were entitlement funds uncontrollable, Boland said, they were also politically untouchable. Thus, the remaining 25 percent of the budget would come under heavy pressure for cuts. In Boland's view, science and technology, which are in the "controllable" 25 percent of the budget, can be squeezed with impunity because they have the weakest lobby. He predicted that science would have to struggle to get even 1 percent of the budget. "That means that the resources available to build the Shuttle-Orbiters . . . will continue to be limited," he said.[5]

As one Apollo lunar landing succeeded another in the early 1970s, there was no overpowering motive in the country to create a new manned spaceflight system. Many space scientists, in fact, opposed it because it would divert funding from unmanned space investigations, as Apollo had done. This view persisted throughout the Shuttle development years. To many eminent space scientists, the Shuttle was bad news.

The attitude toward the Shuttle of a number of them was expressed by Professor James A. Van Allen, University of Iowa, in an editorial in *Science,* the journal of the American Association for the development of Science.[6] The pioneer space investigator whose scientific team discovered the radiation zones that bear his name from *Explorer 1, 3,* and *4* data in 1958–1959 took the position that Shuttle development had sidetracked space science.

"A deep distress is spreading through the community of scientists and engineers who are engaged in space work," he stated. Of 13 satellites placed in orbit by the United States during 1980, he said, only one had a scientific purpose.

"As of 1981, it is almost impossible to obtain a go-ahead for a new scientific mission or for an advanced application mission in space," he said.

Although it was easy to blame this "bleak outlook" on the Reagan administration's policy, he added, it was difficult to argue that expenditures of $6 billion for NASA

and $3 billion for Department of Defense space activities were inadequate.

"It is time to recognize that the dominant element of our predicament is the massive national commitment of the past decade to development of the Space Shuttle and the continuation of manned flight," Van Allen said. It was a commitment, he said, that arose in part "from vaguely perceived future benefits of vast enterprises, such as manufacturing in space, solar power satellites, human colonies in space and mining of the Moon and asteroids." Meanwhile, important space objectives (in science) languished, he said.

Inasmuch as the Shuttle was no longer a unit of an elaborate interplanetary transportation system, NASA found it necessary to justify it as a cost-effective carrier that could lift satellites and probes to low Earth orbit more cheaply than expendable rockets could. The Shuttle thus become detached from the Space Task Group's perception of it as an element in an interplanetary transport infrastructure. It evolved as a system within itself, a space truck, without linkage even to a space station.

In 1970, the Shuttle's potential for opening up a new frontier for industrial expansion seemed no more credible or compelling to its critics than had the prospect of Columbus' opening up a new world in 1492. The Talavera Commission, which examined his claims in 1490, expressed the view that they were dim indeed, and some parallel with critical attitudes toward the Shuttle can be noted in the commission's report to Queen Isabella:

"The committee judged the promises and offers of this mission to be impossible, vain and worthy of rejection; that it is not proper to favor an affair that rested on such weak foundations and which appeared uncertain and impossible to any educated person, however little learning he might have."*

NASA officials quickly discovered that their main hope of winning congressional support for the Shuttle in the

---

*Isabella appointed her confessor, Hernando de Talavera, to head the commission and report to the throne. The version of the report quoted was presented by Dr. Krafft A. Ehricke, chief scientist, Space Division, North American Rockwell, in a statement to the Subcommittee on Manned Space Flight, House Committee on Science and Astronautics, during the 1973 NASA authorization hearings.

atmosphere of indifference and skepticism that was thickening in the early 1970s was to emphasize its reusability as a common carrier and hence its long-term economy in a utilitarian space program. From that perspective, the Shuttle was born out of the context in which it had been conceived, an orphan of the political storm.

If the United States was to continue manned space operations, they would have to be utility oriented, commercially justifiable, and environmentally useful, aside from whatever military justification could be brought to bear. NASA was now faced with the need to rationalize the Shuttle as a cost-effective satellite transport to low Earth orbit. Its role in a grand design for interplanetary travel was summarily dropped, and its potential for developing space manufacturing opportunities became an accompaniment to the main theme of its obvious utility as a reusable space truck.

As a means of survival, manned spaceflight became utility oriented. By the end of 1970, the Shuttle had evolved fully in the context of an orbital delivery system, with wings. Although this perception of it took some of the romance out of spaceflight, it succeeded in getting development money out of Congress.

The Compromise

NASA awarded contracts for a Shuttle Phase B study to McDonnell-Douglas Astronautics Company, St. Louis, and to North American Rockwell Corporation, Downey, California. In Phase B, the concept (Phase A) assumed a physical shape, with dimensions and cost estimates.

Phase B confirmed the juggernaut scale of the vehicle. The booster was the size of a jumbo jet; the orbiter, that of a 707. Payload dropped to 12 tons. In the cabin of the orbiter there were seats for pilot, copilot, and up to 12 passengers.

The preliminary cost of this pterodactyl-sized spaceship as estimated at Phase B was $10 to $12 billion. Ten years earlier, when the United States was being humiliated by Russian space successes and exploding American rockets, the price might have been right. After two

landings on the Moon, the price was way out of line. We still had Apollo and its successful technology. No urgency existed then for a follow-on system.

John F. Yardley, a former McDonnell-Douglas engineer who became NASA Associate Administrator for spaceflight in 1974, recalled the attitude of the Office of Management and Budget. "There is no question that the OMB would have been very hard to sell at $10 to $12 billion," he said in an interivew.[7] "If you looked at the NASA budget level and analyzed it from a structural point of view—how much of that level could you afford to put into your peak funding years without decimating the rest of your agency?"

Yardley and his associates in the agency's top management were convinced that the NASA funding level, which had decreased in terms of purchasing power since the mid-1960s, would never allow the agency to develop anything costing more than $5 billion over a seven- or eight-year period. "Even if the OMB had okayed $8 to $10 billion, NASA would have been taking a hell of a risk," Yardley said. "On a program like this, if you tried to double the time of development, to stretch out the cost [to reduce peak funding], it would just get away from you." He meant that a stretch-out in development would result in higher labor costs, which might become unmanageable.

It became known to the agency, in the way such things become known around Washington (the administrator is usually given to understand by somebody on the White House staff), that $5.5 billion would be top dollar for Shuttle development. The limitation presented the new space transportation system with its first real crisis. Studies had shown quite clearly that the cost of developing the Shuttle would be inversely proportional to the cost of operating it.

A fully reusable Shuttle would be the most costly to build. This was the two-airplane Shuttle, with booster and orbiter capable of being used over and over again. But it was the cheapest to operate. The Phase B studies had indicated that such a Shuttle could place 12.5 tons in a 270-mile orbit at a cost of $2 to $5 million a flight. Orbiting payloads at $80 to $200 a pound would be a breakthrough in space transportation compared with costs ranging up to $1,000 a pound on Atlas, Titan, and Saturn class expendable rockets.

If a fully reusable Shuttle could not be built because of budget constraints, NASA faced the necessity of devising an engineering compromise that would degrade the vehicle to one that was only partially reusable. This would mean abandoning the booster airplane and substituting rockets. The NASA design team explored that fallback position.

It was no use appealing to Congress. Among a number of Democrats in Congress, there was a growing sentiment that the Shuttle was out of context in terms of national priorities. Some of the liberals regarded it as a sop to the aerospace industry and to the military-industrial complex.

If the NASA designers were forced to degrade the Shuttle to a rocket-boosted airplane, with only the airplane fully reusable, would it still be cost effective? That was the question NASA faced as its engineers and design team searched for an alternative design. Desperately they clung to the flyback booster in the hope they could avoid substituting rockets. No one had ever designed a reusable rocket.

Desperation evoked a radical idea. The size of both the booster and orbiter airplanes had been determined largely by their internal propellant tanks. What if the propellant tank for the orbiter were shifted from the inside to the outside? The tank was so much dead weight after engine burnout at the conclusion of the launch to orbit. Why not dump it into a convenient ocean? Of course, the tank could not be reused, but if some part of the reusable vehicle had to go, the tank was the most readily expendable. Replacing it each flight would cost $600,000.

By putting an expendable propellant tank on the outside, the design team could reduce the length of the orbiter from 166 to 122 feet. At the same time, jettisoning the tank would increase cargo capacity by cutting return-trip weight.

The alternative design now consisted of the 747-sized, manned, flyback booster airplane, the 122-foot orbiter, and the expendable tank. That much of a compromise would reduce development cost to $8.2 billion. It would also increase minimum operating cost from $2 million to $3.8 million a flight.

The $8.2-billion development cost was still too high. It could not be reduced unless the flyback booster—the 747-sized airplane—were eliminated. The alternative was to replace it with a rocket, or a pair of rockets. The 156-inch-diameter solid-fuel rocket booster, dating back to 1962, looked like the most effective replacement. Liquid-fuel rockets were also considered, including the Saturn IB.

In order to conserve as much reusability as possible, the designers believed, the booster rockets could be re-

Shuttle design developed by Rockwell International Corp. The artist's drawing shows the orbiter during ascent two minutes after liftoff as the solid rocket boosters are jettisoned. The orbiter continues toward orbit for another six minutes and then drops its external tank, the expendable element of the Shuttle.   (Rockwell)

Three external tanks rest in cradles at NASA's Michoud Assembly Facility outside New Orleans. The tanks were made by the Michoud Division of Martin Marietta Aerospace.   (NASA)

covered, refurbished, and reused up to 20 times. They would fire only for the first two minutes of the launch and then drop off at an altitude of about 27 miles, descending into the ocean on parachutes to be recovered by ship. Although the scheme of recovering the boosters was cumbersome and its savings mostly theoretical, it offered an acceptable compromise.

During the first quarter of 1971, the Shuttle had evolved, or devolved, into the rocket-boosted airplane the advanced systems people at NASA had been hoping to avoid. The Shuttle now consisted of three parts: The air-

plane part, or orbiter, was 122 feet long, with a swept delta wingspan of 78 feet and a cargo bay 60 feet long and 15 feet across. It was approximately the same size as the DC-9 commercial carrier. The second part was the external propellant tank (ET), which carried 518,000 gallons of liquid oxygen and liquid hydrogen to feed the engines in the orbiter. Part three of the Shuttle consisted of twin solid rocket boosters (SRBs), one on each side of the external tank on which the orbiter itself was mounted. They were monsters, too, 156 inches in diameter and 149.5 high, towering 27 feet above the orbiter.

One of the massive tools used to manufacture the tanks—a rotating, expanding mandrel. It is part of an automatic welding machine designed by the Vought Corp., Dallas. The tool aligns, trims, and joins subassemblies to form a tank 99 feet long with the aid of laser beams.   (Vought Corp.)

The transition of the Shuttle from a fully reusable, two-airplane transport to a partly reusable, rocket-boosted airplane is essentially the product of the budget constraints, which reflected priorities for social welfare programs of the early 1970s. A social critic might have observed that the compromise showed that an ambitious technological advance in space could not be fully realized in a period of tumultuous dissent and disaffection.

The enthusiasm for spaceflight which Apollo had generated in the late 1960s collapsed in the early 1970s

under a landslide of social and racial unrest. The shuttle had a small constituency, but it was rejected by many who believed that space technology was dehumanizing. The news media reflected the indifference of the good, gray silent majority to further manned adventure in space.

New Year, 1972

The new year brought decision. On January 5, 1972, President Nixon issued an order from San Clemente instructing NASA to proceed with the rocket-boosted Space Shuttle. "It would," he predicted, "transform the space frontier of the 1970s into familiar territory, easily accessible for human endeavor of the 1980s and 1990s."

This perception of the Shuttle as the Covered Wagon of the High Frontier evoked the tradition of pioneering and expansion that had built the nation. It was a more acceptable motif for spaceflight in the grim 1970s than the romantic adventurism of Apollo.

A new, hard, practical attitude toward spaceflight began to grow, based on the new frontier theme. It was a theme with potential economic substance. Having initially thrust as far as the Moon, the nation retreated to low Earth orbit, a region to become known as "LEO." There it was settling in.

"Our final conclusions," Dale Myers told a news conference in Washington January 10, 1972, "were that we should have a fully reusable orbiter, a hydrogen tank that is discarded in orbit . . . and a choice between recoverable liquid-fuel or solid-fuel booster rockets. The orbiter will be an airplanelike vehicle. It will have . . . payload capacity up to 65,000 pounds."

Myers, a former North American Rockwell executive, was NASA associate administrator for the Shuttle. In his view, the orbiter was essentially a rocket-boosted glider. Its three super-powerful hydrogen-oxygen rocket engines, which were fed by the 154.2-foot-long external tank, were supplemented by twin solid rocket boosters. Developmentally, this was the cheapest form of reusable craft that could lift 65,000 pounds of cargo off the ground. The big booster airplane was replaced by five rocket engines: The three hydrogen-oxygen engines in the orbiter would produce a total thrust of 1.1 million pounds. The two solid rocket boosters would produce a total of 5.8 million pounds of thrust at liftoff. The grand total of 6.9 million pounds came close to the 7.5 million produced by the first stage of the Saturn 5 moon rocket.

For a time, the orbiter design carried a pair of jet engines to give the ship a "go-around" capability in case it missed its landing. These were eliminated, making the vehicle a glider after it reentered the atmosphere.

The final decision on the boosters was to use solid-fuel rockets, based on the 156-inch solid rocket technology of a decade earlier. They would be easier to recover. They appeared to be more economical than liquid-fuel boosters.

The Price

On March 15, 1972, the Space Shuttle cost estimate for design, development, testing, and evaluation of two vehicles was stated to be $5.15 billion. The total cost of manufacturing a fleet of five Shuttles was estimated at $8.1 billion.

On November 6, 1972, Myers advised the House Committee on Science and Astronautics that the first manned orbital flight (FMOF) of the Shuttle was scheduled for March 1978, following the first horizontal flight (glide test) in mid-1977. The Shuttle fleet would become operational in 1979.

NASA projected 581 Shuttle flights between 1980 and 1991 at a cost of $43.1 billion. Compared with similar missions by expendable rockets costing $48.3 billion, the Shuttle would save the taxpayers $5.2 billion—which matched the cost of its development.

NASA emitted a snowstorm of figures, purporting to show how much cheaper it would be to use the Shuttle than rockets. One estimate claimed that the cost per pound to orbit would be $160 on the Shuttle, compared with costs ranging from $900 to $5,600 on expendable rockets.

These claims were regarded with skepticism by the U.S. General Accounting Office, the Congressional watchdog of the federal budget. The GAO was not certain that the Shuttle was economically justified even though NASA's calculations showed that it was, the U.S. Comptroller General reported to Congress June 26, 1973.

NASA executives marched before Senate and House space committees throughout 1972 with optimistic projections. New high-pressure hydrogen-oxygen engines would be built for the orbiter, the most powerful, the most advanced, the most sophisticated in the world. Engine development admittedly would be difficult; not all the technology for the high-pressure turbopump systems was fully developed. But it could be done.

An engine contract was awarded to Rocketdyne, the engine subsidiary of North American Rockwell, which had won the prime contract for the orbiter. The original price of the contract for three test engines and four flight engines was $25,766,000. Five years later, after 895 modifications, the cost had risen to more than $1 billion for twenty-seven engines.

Pursuing reusability wherever it could, the Shuttle directorate determined that whatever else might be expendable on this vehicle, the heat shield should be reusable. During earlier versions of the Shuttle, the designers had seemed to be content with the ablative type of shielding used on Apollo. It was made of epoxy resin and dissipated heat by flaking off.

Epoxy resin was not the ticket for the Shuttle. First, the vehicle was too big; there was too much surface to cover. Second, the material would be too heavy. Finally, it would have to be replaced every flight. The name of the game was reusability.

One of NASA's research laboratories, Ames, at Mountain View, California, proposed a shielding made of silicate fiber—literally, sand. It could be molded into squares six and eight inches on a side and shaped to fit the contour of the orbiter's aluminum skin. In the laboratory, its refusal to conduct heat was remarkable.

The Lockheed Missiles & Space Company at Sunnyvale, California, received the contract to manufacture these squares, or "tiles." It took about 31,000 of them to cover the vehicle. On the nose and wing leading edges, another material, called carbon-carbon (a baked or pyrolized carbon), was to be used. It was produced by the Vought Corporation of Dallas, Texas.

As it was presented to Congress in 1972, there was essentially nothing new about the Shuttle, except the heat shield, or thermal protection system. The engines represented well-known technologies. In space, the orbiter used the engine systems developed on Apollo. Its electric power system (fuel cell batteries) had been evolved in 1962 for Gemini and proved on Apollo. There were no breakthroughs to break through.

After exhaustive hearings, the House Committee on Science and Astronautics drew these conclusions about the Shuttle:

1. The technology and fundamental resources exist to successfully develop the configuration chosen by NASA for the Earth Orbital Shuttle.

2. Within reasonable developmental risk, it appears that the current Shuttle design will allow development total cost to stay within $5.150 billion, as projected by NASA, and allow NASA to achieve an operational cost of $10.5 million per flight.

3. The success in meeting the operational cost per flight for the Shuttle is particularly sensitive to the cost of the propellant tanks and acceptable recoverable or refurbishable cost of the solid rocket boosters.

4. The development of the Space Tug is of key importance in gaining full utility of the Shuttle.

5. If the nation is to realize the full benefits of near space in terms of scientific exploration, practical application, and national security, the development of a low-cost Space Shuttle system is essential.[8]

The new NASA administrator, James C. Fletcher, added his endorsement to this view in a letter to Elmer B. Staats, U.S. Comptroller General, dated May 21, 1973. Fletcher expressed the agency's confidence in the nation's technological readiness to build the Shuttle. "At the time the Shuttle costs were firmed up in early 1972," Fletcher said, "the background of the study, design and engineering was already greater than for most development programs. This started the program off on a sound basis . . . Normally when cost overruns occur in a development, this becomes evident during such a period as we have been through and this has not been the case with the Shuttle."[9]

So ran the record in 1973. Fletcher's confidence in the nation's readiness to build the Shuttle was premature. It was to take seven years before the technological establishment of NASA and the aerospace industry could bring the first Shuttle to the launch pad with a reasonable expectation that it would fly as designed. It would cost the space agency 60 percent more than estimated to develop the Shuttle and three years longer to complete it than planned.

In sum, this "state-of-the-art," "off-the-shelf" project was to become a technological changeling. The Shuttle was to become the most difficult and exasperating engineering challenge of the space age.

The Early Years: The Shuttle Shapes Down

As the Shuttle was being squeezed and pulled into final shape by economic and social pressure, opposition to it mounted in Congress. The Senate Democratic leader, Mike Mansfield (Oregon), announced that he would oppose the $5.15 billion Shuttle investment as a "misplacing of priorities." He was joined by Senators Edmund S. Muskie (D-Maine), Edward M. Kennedy (D-Mass.), and Walter F. Mondale (D-Minn.). Muskie contended that the Shuttle would rob the treasury of badly needed funds for humanitarian projects. He accused Nixon of playing "pork barrel politics."[1] To congressmen outside of California, the Shuttle appeared to be a federal work project for that state's aerospace industry.

Senator Mondale challenged NASA's claim that the Shuttle would reduce tenfold the cost of placing scientific and utility satellites in orbit. He cited a report by the Rand Corporation, a government-financed think tank, denying the claim. "I believe the American taxpayer will conclude that there are more urgent needs here on Earth which deserve priority," Mondale said. He estimated that Shuttle and space station projects would run up a bill of $20 to $25 billion.[2]

During the election campaigns of 1972, the most articulate criticism of the Shuttle seemed to come from congressional liberals. The project seemed to them to be an affront to the poor, unconscionable in a society burdened with social problems.

A curious juxtaposition occurred. Liberals found themselves opposing technical progress, while conservatives supported it.

Among the liberal Democrats in the Senate, however, there was one pro-Shuttle voice. A front-runner in the presidential primary campaign of 1972, Senator Hubert H. Humphrey (D-Minn.), whose liberal record could hardly be challenged, spoke up in favor of the Shuttle as he toured industrial states.

Far from neglecting the hungry, he said, the decision to appropriate $5 billion for the Shuttle would create more than 50,000 jobs in the aerospace and associated industries. As vice president in Lyndon B. Johnson's administration (1965–1969), Humphrey had been chairman of the National Aeronautics and Space Council, an advisory body with a strong influence on space policy. In that capacity, he had gained more insight into the potential of space technology development than most of his colleagues had.

In Chicago, he said that he did not believe that developing space technology and providing for social needs were mutually exclusive. However, many of his liberal colleagues in the Senate and the House, with scant interest in or insight into space programs, supported the Mondale-Muskie-Kennedy-Mansfield viewpoint. In this period of gestation of the Shuttle, the notion that space activity diverted funds from human needs influenced the final decision to build the Shuttle on the cheap. The more NASA sought to emphasize the cost-effectiveness benefit of the Shuttle, the more convinced its opponents became that it was a make-work scheme of the military-aerospace complex. Whatever long-range potential a space transportation system might have for human society was lost in the shuffle of immediate priorities.

The argument that space projects took bread out of the mouths of the poor was dramatized for the first time on the eve of the launch of *Apollo 11* to the Moon.

As crowds gathered on the Florida beaches for the launch, a march of poor blacks and a mule train led by Dr. Ralph Abernathy arrived at the Titusville gate of the Kennedy Space Center. Its mission was to protest the firing off into space of millions of dollars that the poor needed for food, clothing, and shelter. The space agency handled the demonstration by transforming the demonstrators into spectators. Food and shelter were made ready when the sixty men and women of the caravan, who had walked all the way from Georgia, arrived at dawn. Everyone was escorted to the VIP viewing area to watch the launch. The protest was thus absorbed by the event.

Unmanned scientific programs in space also became targets of social critics. A "science for the people" movement invaded the 1970 meeting of the American Association for the Advancement of Science in Chicago to express a critical view. One of its demonstrations interrupted a lecture on the findings of *Mariner 9* at Mars by William Pickering, director of NASA's Jet Propulsion Laboratory at Pasadena. The *Mariner 9* discoveries were epochal in the history of science, but a gaggle of youthful protesters was not concerned with that. They marched across the platform in front of Pickering carrying signs that read, "West Madison St.—Not Mars!"

West Madison Street is, or was, Chicago's Skid Row. The message of the protest seemed to be that the scientific exploration of Mars interfered with the rehabilitation of drunks, drug addicts, and derelicts on Skid Row. In some manner, *Mariner 9* had become the symbol of Chicago's social disabilities.

Skid Row is gone from West Madison Street now. It used to be more of psychopathic sink than a social one. Its denizens were not merely the elderly poor, frequently depicted as clutching a gin bottle in one hand and food stamps in the other, but mostly the neglected mentally ill. The protesters—many of them teenagers from comfortable North Shore homes—would have shown better aim if they had picketed the Illinois Department of Mental Health.

The Consultants' War

If the Shuttle debate had any impact on the election of 1972, it was not measured. Nixon, who had firmly supported the project, was reelected overwhelmingly. Either the Shuttle was inconsequential as an issue or the silent majority was generally for it.

As Humphrey predicted, however, the economic effects of the project were strongly regional. On July 26, 1972, North American Rockwell's Space Division at Downey, California, received the prime contract for design, development, and manufacture of the orbiter and for its integration with the solid rocket boosters and the external tank. Rockwell people staged an eight-hour celebration at a nearby restaurant that had been the scene of a similar victory rite for the Apollo prime contract a decade earlier. Rockwell announced that it expected to engage 10,000 subcontractors and suppliers. The estimated value of the contract was $2.6 billion.

Rockwell won out over Lockheed Missiles & Space Company of Sunnyvale, California, McDonnell-Douglas Astronautics, St. Louis, and Grumman Aerospace of Bethpage, New York. The *New York Times* reported (July 27, 1972) that 1,500 Grumman workers faced layoff as a result of the Rockwell award.

With the beginning of the development stage, the question of Shuttle economics came up for debate. As mentioned earlier, the General Accounting Office did not believe NASA's claim that the Shuttle would save enough money in launch costs over a 580-mission period to pay for itself. NASA's case was weakened by the compromises that made the Shuttle only partially reusable. The expendable propellant tank added $600,000 to each mission in operating costs, and the system for recovering and refurbishing the solid rocket boosters was calculated to amount to one-third of their cost. The recovery required two ships, an elaborate parachute deployment for each of the SRBs, and a device called a "nozzle plug," which alone cost about $1 million for each rocket.

Moreover, although it was assumed that the heat shield would be reusable, there was no guarantee. The shielding was not tested through an actual reentry. As development proceeded, the mechanical strength and adherence qualities of the silicate tiles which formed the bulk of the shielding became more and more doubtful. It was a virtual certainty that part of the shielding would have to be replaced every flight.

NASA hired a consultant, Mathematica, Inc., of Princeton, N.J., to make an assessment of Shuttle costs over a 12-year period, 1979–1990. Extrapolating data from Mathematica's report and from other sources, NASA issued a report in October 1972 stating that a savings of $13.4 billion would be realized in 580 projected NASA and military launches by using the Shuttle instead of expendable Delta, Atlas-Centaur, or Titan 3 rockets.[3]

Of the $13.4 billion, $5.1 billion would be saved on actual launch costs and $8.3 billion on payload design costs. For example, in the enclosed cargo bay of the orbiter, satellites would not need the aerodynamic protection of shrouds or the complicated, often troublesome mechanism to release them in space.

With the $5.15-billion development cost of a flight article and four more orbiters to comprise a five-Shuttle fleet, the projected cost of the Space Shuttle Transportation System was reckoned by NASA at $8.04 billion.

With these data, a magical feat of subtraction could be performed. By deducting the $8.04-billion cost of the fleet from the $13.4-billion savings in launch and payload costs to be realized from it, the space agency could show a "profit" of $5.36 billion over the 12-year operating period. With a full payload of 65,000 pounds, NASA figured that it could haul payloads to low Earth orbit for $160 a pound, compared to the $600, $700, or $1,000 a pound charged customers using expendable rockets.

The assessment was speedily attacked. One critic was the well-known physicist and nuclear power consultant, Ralph Lapp. He appeared before the Senate Committee on Aeronautics and Space Science on April 12, 1972, to challenge NASA's claims. Lapp explained that he supported a general movement among scientists to challenge new technical proposals as a means of requiring their sponsors to prove their worth. Using NASA's and Mathematica's calculations, Lapp argued that instead of transporting material to orbit at $160 a pound, the Shuttle would incur a transit cost of $5,100 a pound, or close to it.

Lapp argued that NASA's cost estimate was based on an operating cost of $10.5 million per flight with a full payload of 65,000 pounds. However, when total costs of the space transportation system, including research, development, testing, and engineering, are taken into account, NASA's figures show a cost of $25.5 million per flight for 514 Shuttle flights, he said. Moreover, instead of 65,000 pounds, cost per pound should be based on average loading of the cargo bay, which Mathematica had estimated at 5,000 pounds per flight.

"If we divide $25.5 million by 5000 pounds, we get a unit price of $5,100 per pound placed in orbit," Lapp said. "This I maintain is the true price since it is based on a full accounting of new appropriated dollars for the space program transportation system. . . ."

He added that Shuttle proponents may argue that their price is based only on operating costs, but "that isn't the way General Motors figures it when I buy a car. . . ."[4]

Mathematica's director, Klaus P. Heiss, told the committee that Lapp's cost analysis had no practical significance. It was derived from NASA's "passion for oversimplification," he said. Numbers relating to a cost reduction on a per-pound basis were unreal because they were calculated on the Shuttle's payload capacity on a due east launch, he said. Aided by the Earth's rotation, a due east launch allowed the orbiter to carry more weight than a polar launch. Many payloads would not be launched due east, Heiss pointed out, and therefore would have to be less than 65,000 pounds. Some would be launched into polar orbits from Vandenberg Air Force Base, California. They would head due south over the Pacific to fly over the south pole. Also, Heiss pointed out, NASA's figures were based on a 100-nautical-mile orbit. Many payloads would be boosted higher than that, requiring more energy, and consequently would cost more.

Heiss said, "NASA's passion for this oversimplification has left it vulnerable to attack by uninformed critics." The only valid way to compare launch costs of the Shuttle with those of expendable rockets, he said, was to calculate them in terms of actual payload. He presented a hypothetical example based on launching a large satellite called *Radio Explorer*.

The cost of preparing it for launch to low Earth orbit in the Shuttle was figured at $222 million. The cost of preparing it for launch aboard an expendable rocket was $305 million—$83 million more. Actual launch cost would be $103 million by expendable rocket, but only $54 million by Shuttle, a savings of $49 million. Total savings using the Shuttle—$132 million.[5]

Dead-Stick Landing

History will show whether Lapp or Heiss had the better of the argument. Lapp was right about one thing: the technological risk of the Shuttle was higher than anyone in NASA believed in 1972—and that would have a bearing on its development and operational cost.

One factor was weight. It has been the bugaboo of every manned space vehicle ever developed. They all get fatter than planned. The *Enterprise* did, to the point where it was eventually relegated to the role of a test and exhibition vehicle. *Columbia,* the second development orbiter, was constantly in jeopardy of getting overweight. Every pound of excess weight meant one less pound of cargo.

The struggle to keep down weight and cost resulted in late design changes in the orbiter. The major one was the elimination of jet engines that would have made it possible for the crew to recover from a missed landing approach. This departure from the 1972 design reduced the orbiter's maneuverability in the atmosphere to that of a glider. I do not recall that the change was widely heralded when it was decided, late in 1972. Some members of Congress heard about it for the first time, it appears, during a NASA appropriations hearing March 6, 1973, before the Senate Committee on Aeronautics and Space Science. Senator Barry Goldwater (R-Nev.) asked how much fuel the orbiter would carry for maneuvering in the atmosphere after reentry.

None, came the surprising answer. "We have made a decision to do a dead-stick landing from space," explained Dale Myers.*

During 1972, NASA had experimented with high-speed gliding—"high-energy approaches." The energy was gravity. If the orbiter could be landed as a glider, the cost and weight of "air breathing" jet engines, tanks, and jet fuel could be eliminated.

At reentry, the orbiter would have enormous kinetic energy, imparted to it first by the launch and second by its fall around the Earth. Translated in terms of velocity, it amounted to about 5 miles a second. That's 18,000 miles an hour. The space plane had to slow down to 300 mph to land. So most of the energy was transformed into heat as air friction slowed down the vehicle. The heat

---

*Myers told a news conference in Washington March 15, 1972, that the orbiter was being designed to land either with or without turbojets. He said, "We can develop an automatic landing system for what amounts to a dead-stick reentry from space."

shielding deflected the heat; otherwise the orbiter would burn like a meteorite.

From reentry on, the velocity of the craft was managed by a delicate balance between lift and drag forces in the atmosphere and by the gravitational pull of the Earth. The balance controlled the energy of the descent. It required a superhuman pilot; it required a computer, or, as in the case of the orbiter, a committee of them.

The jet engines had been considered only to give the pilot go-around capability in case he missed his correct final approach. But advances in flight computers and the development of a new electronic landing system made the landing look virtually foolproof.

During the early years of Shuttle development, it was assumed that after being launched from the Kennedy Space Center (KSC) on the east coast of Florida, the orbiter would land there. The flight was tentatively mapped as follows: Launched at an inclination of 28.5° to the equator, or due east from KSC, the spaceship would reach an orbit of 185 miles or so over the Indian Ocean and fly a mission of two to nine days.

A tentative flight plan, projected at my request by Mission Control at the Johnson Space Center, Houston, indicated that the pilot would commence deorbit (descent from orbit) over the Indian Ocean as the ship approached Australia. This would be done by firing the orbital maneuvering engines in the direction of flight. Following the retrofire maneuver, the orbiter would reenter the atmosphere, nose up, southwest of Hawaii and glide across the Pacific, emerging from radio blackout as it approached Mexico.

Crossing the Gulf of Mexico, the space plane would pass the west coast of Florida between Tampa and Sarasota and begin its descent into the KSC's three-mile-long runway east of Orlando. Except for a window-rattling sonic boom or two, its passage would be silent. Its landing speed was expected to be no more than 260 miles an hour.

Thus, the orbiter would have glided about one-fifth of the way around the planet, dead stick, with no source of power save that which moves the planets around the Sun and the Solar System around the center of the galaxy—gravitation—plus the energy imparted to the spaceship by the launch.

The scenario was a new one in the annals of flight. The only conventional aspect of it was the flight plan NASA would have to file with the Federal Aviation Administration Center in Miami. In the atmosphere, the orbiter is definitely an airplane and subject to air traffic control.

Such was the nominal flight envisioned by the Shuttle team. The reality would be somewhat different, but no less of an astronautical tour de force.

During the summer of 1972, NASA test pilots made a series of dead-stick landing tests with a Convair 990 aircraft at the agency's Dryden Flight Research Center in the Mojave Desert. The tests were designed to simulate the descent of the orbiter without power from 40,000 feet. From an imaginary "gate" at 20,000 feet, about 20 miles from the touchdown point, the Convair settled into a steep glide. Its descent velocity was about 345 miles an hour. Then, 45 seconds from touchdown on the Rogers Dry Lake runway, the pilot or the autopilot pulled up the nose, causing the craft to flare, or level off. With a second flare, the 990 landed quietly. The tests were flown with pilot and with autopilot.

Myers told the committee that the 990 tests had demonstrated the feasibility of an unpowered approach and landing by the orbiter. It was on this basis that the turbojets were eliminated. A "foolproof" landing system would compensate for the go-around capability. The landing would be controlled by the orbiter's digital flight control system, responding to the guidance of a microwave scanning beam on the ground. The beam provided an invisible slideway along which the heavy glider descended under computer control. The human pilot would monitor the automatic landing system.

With the dead-stick decision, NASA cancelled contracts for the purchase of Pratt & Whitney C-141 jet engines. The agency also purchased a second-hand Boeing 747 jumbo jet from American Airlines for $15.6 million in 1974 and converted it into a carrier for the orbiter at an additional cost of $35 million. It was this aircraft that carried the *Enterprise* aloft for the approach-and-landing tests of 1977.

At the Senate hearing March 6, 1973, Goldwater, who has experience as a private aircraft pilot, expressed surprise at the dead-stick landing decision. It was quite a benchmark in aviation history. Was it safe? Reliable?

"We have repeated landings [of the 990] from 40,000 feet with zero power at the sink speed of less than two feet a second at touchdown," Myers assured him. "We are convinced that this is the proper way to handle this job, rather than with space-rated turbojets, which is the alternative."

Goldwater: You also have the pilot convinced out there that 300 knots (345 miles per hour) is a good approach speed?

Myers: Yes.

Goldwater: If you have a 12-mile runway, that may be all right.

Myers: We have plenty of margin in our capability to stretch or decrease that glide slope as we come in. When we are down to subsonic speeds, we are within about a two-mile accuracy relative to the landing site. From that point on, the drag variations we can handle through angle of attack will surely take care of that. At the Kennedy Space Center, we would be subsonic and at 300 knots at 40,000 feet. We duplicated that on the last part of the run with the Convair 990.[6]

The dead-stick landing made a profound change in the entire flight profile of the orbiter. It introduced an element of uncertainty that would not be resolved until the first orbital flight was made, despite the 990 tests.

Turbojet propulsion had been accepted hitherto in the final stages of the descent from orbit. It would have enabled the orbiter to reenter the atmosphere at a 60° angle of attack (nose up 60° relative to the ground). The orbiter would have maintained this nose-up attitude until it reached 40,000 feet, where its velocity would have dropped to 300 feet a second (204.5 mph). Then the nose would have been pushed down and the orbiter would have dived until it gained sufficient velocity to glide in level flight. On final approach, the pilot would have started the turbojet engines and made a normal aircraft letdown and landing.

Removal of the turbojets changed the descent strategy. Without the jet engines, the orbiter would reenter at a 40° angle of attack. Its glide path would be longer and its velocity higher than if it had reentered at 60°. But without the engines, the vehicle had to conserve energy to maneuver for the landing.

One effect of these changes was a reduction in the heat load (intensity) as the vehicle reentered the atmosphere. Another was a greater reliance on the precision of the multicomputer flight control system.

The orbiter would have the most sophisticated flight control system ever developed. But like the heat shield, its effectiveness could not be determined in advance of an actual flight test.

No Escape

Another change in the original scheme broke the long-standing tradition of providing emergency escape for the crew during launch. Solid-fuel escape rockets had been considered, to lift the orbiter off the pad, away from the external tank and the solid rocket boosters, so that it could glide off to safety in the event of an impending blowup. In theory, the rockets would lift the orbiter high enough to allow it to land on the Kennedy Space Center runway. If not, the crew would have to be rescued at sea. The orbiter would float unless it were carrying a full 65,000-pound payload—in which case its flotation capability was uncertain.

Although an explosion on the pad was put down as an unlikely possibility, of interest only to sensation-seeking writers, the Shuttle safety team had considered it quite seriously. The 154-foot-tall propellant tank was a potential bomb when loaded with 378,000 gallons of liquid hydrogen and 139,623 gallons of liquid oxygen. NASA's 1977 Environmental Impact Statement calculated the maximum explosive force of the propellant as equal to that of 6.3 million pounds of dynamite. It said that the blast wave would be lethal for 1,000 feet; produce lung injury at as far as 1,700 feet; rupture eardrums at 2,000 feet; hurl penetrating missiles 5,000 feet; and break windows at 13,000 feet. If the tank blew up, it would take the solid rocket boosters with it—and, of course, the orbiter. Burning propellant in the boosters would emit a cloud of poison gas, composed mainly of chlorine and hydrogen chloride.

During a nominal launch, the escape rockets would be dropped as soon as the orbiter left the pad.

Escape systems had been provided on all previous

manned spacecraft. Mercury and Apollo had steeple-shaped escape rockets that could pull the capsules away from the launch vehicles to an altitude from which they could descend safely by parachutes into the sea. Gemini spacecraft had aircraft-style ejection seats from which the crew could be hurled out of the vehicle and parachute to safety.

As modified in 1972, the orbiter was the first American space vehicle without a launch escape system as an element of basic design. Ejection seats were provided in *Columbia* for the first four test flights. They would exit the cockpit through an overhead hatchway, which would be blown out of the top of the cabin by explosives.

This escape system was temporary and limited. It could not be used while the orbiter was on the launch pad, because the ejection path (or vector) would be parallel to the ground and much too low to permit a parachute descent. After the four test flights, the ejection seats were to be removed, and there would be no escape from the vehicle once the countdown reached the automatic sequencer, a few seconds before ignition of the rockets.

However, escape would be possible in the event of an impending blowup if there were still time to stop the countdown. The crew and the small band of "close-out" technicians remaining on the service tower could escape by sliding down 1,200-foot steel cables in baskets from the 147-foot level of the service structure to the ground. They could then take cover in blast-proof bunkers. The system, installed during Apollo, had never been used in an emergency.

The cockpit fire that asphyxiated the prime crew of *Apollo 1* during a routine flight rehearsal on Launch Pad 34 January 27, 1967, was the only fatal accident involving a U.S. space crew. It was the result of an electrical short circuit in the command module that set fire to plastics in the high-pressure oxygen atmosphere of the cabin, a horrendous example of oversight.

Shuttle "Firsts"

For the first time, Shuttle operations would require firing hydrogen-oxygen engines at ground level. This had not been done before because hydrogen engines had been confined to upper stages of the Atlas (Centaur) and Saturn (S-II and S-IVB stages). The degree of hazard they posed, in the proximity of the external tank, was moot. NASA people simply did not talk about it.

There would be other "firsts" in Shuttle operations. Never before had a winged vehicle encountered the forces of maximum dynamic pressure during launch through the atmosphere. Wind tunnel and computer data indicated that the orbiter would not be seriously affected; besides, its main engines were throttleable and its velocity could be reduced if vibration appeared threatening. But the whole story would not be known until the Shuttle was actually launched into orbit.

Never before had a winged spaceship reentered the atmosphere. Never before had a 75-ton glider descended through the entire regime of hypersonic and supersonic speeds to make a pinpoint landing 5,000 miles from the point of reentry.

Never before had a manned craft undergone its first orbital flight test with a live crew.

It is no surprise that the safest part of the orbiter's flight regime would be the orbital part. There, the airplane would function as a spacecraft, and a spacecraft's behavior in that medium is more predictable than a glider's behavior as it descends rapidly through the high atmosphere.

Like Apollo, the orbiter would have two sets of space engines, both sets highly developed and seemingly mature. Mounted on the aft fuselage, near the big triad of rocket engines, would be two orbital maneuvering engines capable of 6,000 pounds of thrust each. They would provide the means of changing altitude and of braking velocity for reentry. Attitude control, in space and in the atmosphere, would be provided by 44 small reaction control systems (RCS) thrusters.

After launch, the next hazardous segment of the flight would be reentry and descent. Would the heat shield hold? Its thermal insulation quality had been tested in the laboratory, but its mechanical integrity as part of the orbiter structure could be verified only in the actual reentry.

NASA officials were asked time and again why they did not test the heat shield on a small reentry vehicle. The common denominator of their answers is that they did not believe it was necessary to spend the money on such a test. Had they done so, they might have saved millions of dollars spent on reworking the heat shield and a year or more of delay.

Sending a live crew on the first orbital test confirmed NASA's perception of the Shuttle as a space-adapted airplane, rather than as a spaceship capable of controlled gliding flight in the atmosphere. From that viewpoint, using test pilots followed aircraft test tradition.

Flight Instability

During hearings before the Subcommittee on NASA Oversight of the House Committee on Science and Astronautics in February 1974, the flight quality of the orbiter was reviewed by Christopher C. Kraft, director of the Johnson Space Center at Houston.

The final design of the orbiter left unresolved questions about its stability at supersonic velocities. Instabilities in the vehicle's flight behavior had shown up in wind tunnel data. The design team believed that the problem could be handled by the flight computers, which could react to correct an unstable condition faster than a human pilot could. Still, committee members were concerned. The hazard imposed by instability at high Mach numbers (multiples of the speed of sound) would not be entirely known before the actual test flight.

Kraft, an aeronautical engineer whose career spanned the transition of NACA to NASA, had been a sure-footed flight director during the early days of Mercury and Gemini. During critical times in the flights of those vehicles, he had been called on to make some tough decisions. They had turned out to be right.

It was Kraft, along with Robert R. Gilruth and the late Wernher von Braun, who masterminded the spectacular flight of *Apollo 8* to lunar orbit in December 1968. Kraft succeeded Gilruth as director of the Center at Houston when Gilruth retired in 1975, and projected into the Shuttle era the confidence in engineering design that had led the directorate to take calculated risks in Apollo.

So far as flight operations went, Kraft was one of the key members of the great team that had built and operated the Apollo-Saturn system. Von Braun and other ex-patriate German engineers designed the Saturn rocket system at the Marshall Space Flight Center in Alabama; Maxime Faget and his colleagues at the Langley Center in Virginia designed the Apollo spacecraft. Gilruth and Kraft directed the lunar missions. Kraft assured the House subcommittee that the instabilities the orbiter might experience during its long glide home would present no serious risk.

Some members of the design team had suggested that a ventral fin on the rear end of the orbiter would help stabilize it at high speeds. But it would also cause heating and possibly control problems at reentry, Kraft told the subcommittee.

"A lot of our astronauts have been very unhappy and outspoken about the lack of good flying qualities, particularly in the lateral directional modes at Mach numbers from about eight to five, or down to three maybe. We've made a decision to use the RCS system [the small gas thrusters] as a force-producing device, rather than a ventral [fin]," he said.

Instead of using the big rudder and the elevons, which are movable surfaces of the delta wing, to control the orbiter at ultrahigh speeds, the flight computer would fire the small gas thrusters in the tail for steering, just as it would in orbit. At high Mach numbers, the rudder and elevons simply would not produce the results they would in the lower atmosphere at lower speeds.

To yaw (turn right or left) by using the rudder might cause the orbiter to roll or sideslip, Kraft said. Instead, the vehicle could be turned by using the aft jets. At some point in the descent, the rudder could be brought into play along with the jets. But the rudder would not be used alone until velocity fell below Mach 2, twice the speed of sound.

The pilots were not happy with this. Kraft said that in his 29 years of experience with test pilots, he had found that they invariably want inherent dynamic stability in anything they fly. But they weren't going to get it in the orbiter.

"If we're going to fly a Shuttle in the 1970 time period, we're not going to build an aerodynamically stable vehicle," Kraft said.* "That's an impossibility. We're building

---

*In 1974, NASA was confident the Shuttle would fly in 1978 or 1979.

a fly-by-wire vehicle at present that's going to get its stability from black boxes [computers]. And that's the name of the game in the Shuttle."

Fly-by-wire was a control system installed on the Mercury spacecraft. By moving the stick or hand controller, the pilot generated an electrical signal which told a "black box" to fire a space engine or attitude control thruster system. Similar but more complex systems were built into Gemini and Apollo spacecraft.

The orbiter would be even more complex because it would have not only space engines but also aerodynamic controls. In the high atmosphere, both systems would have to work together. Between the pilot and the systems that actuate the aerodynamic and thruster controls would be a committee of computers, functioning as the most complex autopilot ever built. They would make it work.

Kraft admitted that landing the heavy glider was going to be difficult, but he added, "If you're going to fly this vehicle into space and get it back and reuse it, we don't see any other way to do that. You just end up with a low lift-to-drag vehicle that comes down like a rock. However, we're confident that with the avionic [flight control] systems that we have on board and with the judgment of the pilot on top of that, we can build a system that is flyable with proper training."[7]

Design by Committee

The giraffe has been cited as an example of a creature that was designed by a committee, but for all of its gawkiness, it is a survivor species in its environment. Presumably the "committee" knew what it was doing.

The Shuttle, too, was designed by a committee, one on a somewhat lower plane of existence. Not all of its members, lacking omniscience, agreed on everything. Besides, one of the principal influences in the design of the Shuttle was the national budget and its keeper, the Office of Management and Budget.

The giraffe works, but would the Shuttle? One of the principal designers of the orbiter was Maxime Faget, who designed Mercury. NASA inherited Faget from NACA, along with Gilruth, Kraft, and other aeronautical spe-

cialists in 1959. Faget's office was moved from Langley Research Center at Hampton, Virginia, to Houston in 1965 when the Manned Spacecraft Center, later the Johnson Space Center, materialized on the Gulf plain southeast of Houston as the control center for manned spaceflight.

Faget explained in an interview why the orbiter is not stable at hypersonic speed. A vertical fin near the orbiter's far-rear center of gravity would have improved its stability but reduced its cross-range flight capability. Cross range is the distance a vehicle can fly to either side of its orbital flight path.

Faget said the Department of Defense insisted on a cross range of 1,100 nautical miles (1,265 statute miles). In an emergency, the orbiter could land just about anywhere in North America, Western Europe, or North Africa where there was a 15,000-foot runway. Faget agreed with Kraft that the fin would have offered a heating problem at reentry.

Even without the fin, the vehicle's ability to fly to either side of its descent path would probably be limited to 900 nautical miles (1,035 statute miles), according to Faget. Because the rudder was placed above the vehicle's center of gravity, the pilot cannot use it to make a turn at high speed and altitude. If he tries, he will get a roll instead of a yaw, Faget said. Also, the center of gravity of the orbiter is so far to the rear that the pilot would have to use "a lot of rudder" to make a turn, even at a moderate speed in the lower atmosphere. "At Mach 2, the rudder is practically useless," he said. "And we don't use it. We're using the aft thrusters right down to Mach 2 for steering."

The forward thrusters are used in space, but not in the atmosphere. Faget explained that when they fire, they create aerodynamic interference which can cancel or reverse their effect. The pilot might fire them to control pitch and get a yaw out of it. But the rear thrusters are free of this cross-coupling effect, and the vehicle can be steered adequately with any two of them on either side, he said.

The elevons, which function as ailerons do on conventional airplanes, would be used for pitch control and roll in conjunction with pitch and roll thrusters.

Thrusters are used during the descent until an accelerometer senses sufficient aerodynamic force to make the elevons and rudder effective. It then notifies the flight computer, which activates these surfaces and blends them in with the thrusters, eventually turning off the thrusters as the aerodynamic force builds up.

I talked with Faget in Houston October 27, 1977, the day after the final glide test of the *Enterprise*. The horrendous engine and heat shield problems were just beginning to appear, but their magnitude was then unsuspected. The engineers and astronauts at Houston were concerned about Fred Haise's rough, bouncy landing, and there were intermittent conferences throughout the day to decide whether the final glide test should be repeated. It was not, and when that decision was reached, everyone seemed to relax.

The 60-year-old Max Faget belonged to the first-generation spaceflight design team. It was the team that had conjured up manned vehicles from the Mercury capsule to the reusable winged spaceship. Wiry, intense, with a fringe of gray hair and bright blue eyes in a tanned, crinkled face, Faget talked in a low, soft voice, almost a growl. He was not entirely satisfied with what he and others had wrought, but he was confident it would work.

"We've as much wind tunnel [experience] on this one as on any machine ever put forward for flight," he said. "We've put an awful lot of aeroscience manpower in analysis of that data. We are also analyzing the uncertainties, although we have a fairly good understanding of what to expect.

"You see, with a conventional airplane, you can always drop back from unknown flying conditions to known flying conditions if you get into trouble. With the orbiter, it's the reverse. On reentry, we begin with the unknown and hopefully reach the known. We start off at the highest speed and we're absolutely committed at that time to fly through the complete flight regime. There's no way we can back out of any condition.

"This design is the melding of a great number of considerations. Probably those put forward by 30 or 40 individuals. If I'd had my druthers it would not have turned out this way to begin with. But I don't think you could find any one individual that would have said, well, that's exactly what I'd like to see."

It was the military desire for extensive cross range that led to the delta wing.

"I wouldn't have put that much cross range in it," he said. "I may have used a wing in the rear, and then I'd put a canard [short wing] in front."

Faget believed that the orbiter should reenter at a higher angle of attack than 40°, but that would limit the cross range that the military wanted.

About the orbiter's performance on the glide tests, Faget said he was satisfied. The wind-tunnel data didn't lie, he said.

"How it will do in the high atmosphere—we'll have to see about that when the time comes."

4

The Gold over the Rainbow

Historians identify three phases in the development of the New World: reconnaissance, exploitation, colonization. This pattern reappears in the space age on an abbreviated scale. Reconnaissance was overtaken by exploitation before we landed on the Moon.

By 1968, communications satellites had demonstrated a commercial value far exceeding the cost of their development and launching. They provided the first return on the American investment in space technology.

There was gold in space near the Earth. Meteorological and ground observation and monitoring satellites had proved of incalculable value to agriculture, forestry, air and sea navigation, and flood and hurricane warning systems. The reconnaissance of the Moon in Apollo disclosed the existence of mineral resources that might be useful in developing solar power satellites more cheaply than they could be built from terrestrial materials.

By 1970, it had become apparent to space scientists and technologists that near-Earth space offered a new

frontier for industrial investment and expansion, especially in the realm of electronics, pharmaceuticals, and high-strength alloys.

The two advantages of space processing—high vacuum and low gravity—were unobtainable on the ground. But their utilization depended on cost-effective transport and laboratory facilities.

In the early 1970s, the Shuttle program promised the transport, but the laboratory facility was lacking. *Skylab,* the bungalow-sized space station NASA put in orbit in 1973 and abandoned in 1974, had demonstrated the prospects of such a facility for industrial as well as scientific experiments. However, Congress was not inclined to follow it up. Consequently, the space station that had been linked to the Shuttle in the Space Task Group Report of 1969 was not funded. This deficiency restricted the Shuttle to the role of common carrier for satellites, in lieu of expendable rockets.

When it became obvious in 1971 that Congress would

not buy a space station and a Shuttle in the same decade, engineers and designers at the Marshall Space Flight Center at Huntsville, Alabama, proposed a manned laboratory that could be carried aloft as payload in the 60-foot cargo bay of the Shuttle-Orbiter and returned with it to the ground. This was a variation on an old idea. During the 1960s, the Air Force had invested $200 million in a Manned Orbital Laboratory (MOL) project. The MOL was an aluminum can 10 feet in diameter and 18 feet long—on the order of the Salyut space stations the Russians developed and were using successfully during the 1970s. It was to be placed in orbit by a Titan 3-C rocket and visited by astronauts in a Gemini B spacecraft. An elite corps of Air Force astronauts was organized and drilled for the project, but the Department of Defense was compelled to abandon it. No military missions of sufficient urgency to justify its $3 billion estimated cost could be found. Seven MOL astronauts transferred to NASA.

The Marshall design simply called for an MOL-type laboratory to be hauled into orbit by the Shuttle. The laboratory was called RAM, meaning research and applications module. It consisted of a pressurized cylinder plus an open platform or pallet on which scientific instruments could be mounted. It would go aloft but, unlike the MOL, would return with the Shuttle. Later, the cylinder and pallet array could be modified as free-flying units and left in orbit by the Shuttle, to be picked up on another mission. Marshall awarded a contract to General Dynamics, Convair Division, San Diego, for a Phase A feasibility study.

The study moved swiftly to Phase B, the design stage. Three aerospace industrial teams, including seven European firms, submitted proposals early in 1971 for the project definition studies and preliminary designs.

General Dynamics won the Phase B contract, worth about $2 million. Its team included North American Rockwell, TRW, and Bendix as major subcontractors. Working with them were ERNO, a division of VFW-Fokker in West Germany; SAAB, Sweden; MBB, West Germany; Selenia, Italy; and MATRA, France.

RAM was to be 14 feet in diameter and consist of attachable modules and pallets up to 52 feet in length. Its maximum weight would be 32,000 pounds, which was the estimated maximum the orbiter could return safely to Earth.

The pressurized laboratory evolved in three units: a support module that would house the experimenters as well as power, environmental control, and data-handling equipment; a laboratory module where experimenters could work in shirt sleeves; and an open pallet. This array could either remain in the cargo bay during flight or be swung over the side on hinges for better access to space. Colloquially, RAM became known as the ''Sortie Can'' or ''Sortie Lab.''

To the great disappointment of the Marshall team and its allies, the project did not survive Phase B. The Office of Management and Budget said no. NASA's funding prospects would not support it and the Shuttle, too. Not at the same time.

Would Western Europe consider building RAM for the Shuttle? In 1970, as I mentioned in chapter 2, Tom Paine had sought to arouse European interest in building the space tug, which Congress also had put on the shelf. But European failures in developing the Europa heavy launch vehicle through the European Launcher Development Organization (ELDO) discouraged the project.

RAM was another matter. It fitted neatly into European aircraft technology. A reorganized European space program took up the project and agreed to build a flight unit at European expense for a piece of the action in space. In this way, a partnership was forged that brought finan-

SHUTTLE CAN ACT AS
SHORT-DURATION
SPACE STATION

Artist's concept of the Shuttle circa 1970.   (NASA)

cial and industrial resources of the Western Alliance into the Shuttle program. RAM then became known as "Spacelab."

So it was that Spacelab, conceived in north Alabama, underwent its gestation and birth in a factory in north Germany.

Transorbit Gaul

As Julius Caesar observed in 49 B.C., "Omnia Gallia in tres partes divisa est," which, as every high school student used to know before Latin got lost from the curriculum, means "All Gaul is divided into three parts." In terms of space technology, it still is. There are the Federal Republic of Germany, which led the development of Spacelab; France, which led the development of the first operational European launch vehicle, Ariane; and Belgium and the Netherlands, which have specialized in satellite research and development.

They form the core of the 11-nation West European counterpart of NASA, the European Space Agency (ESA). It evolved from a consortium created by a merger of ELDO and the European Space Research Organization (ESRO). Other members are Denmark, Ireland, Italy, Spain, Sweden, Switzerland, and the United Kingdom. Austria and Norway are associate members, and Canada has observer status. The 11 full members contribute one-tenth of one percent of their gross national product to the ESA budget.

The West European consortium made an agreement with NASA in August 1973 to build Spacelab, in return for which NASA would purchase a second flight unit from the consortium. The first flight article would be used jointly by European and American scientific organizations on Shuttle flights.

When the European Space Agency was formally activated April 15, 1975, the development of Spacelab was placed under a centralized management authority. NASA representatives were assigned to it to assist in integrating the construction of Spacelab with that of the orbiter.

Initially, the cost of Spacelab was estimated at 380 million ESA accounting units, which amounted to $494.7 million in 1975 U.S. dollars. It was a fortuitous deal for both sides. The European consortium acquired access to space. It could send qualified scientists on Shuttle missions as civilian or payload specialists. For the United States, the prospect of European collaboration lent badly needed support to the Shuttle, impressing on a divided Congress that the project now was of international significance. Although the Shuttle might have survived without Spacelab, the deal with ESA made it difficult indeed for any cost-conscious administration to scuttle the Shuttle. It may be that the NASA-ESA partnership, along with Department of Defense interest, kept the Shuttle on track when its development ran into engine and heat shield difficulties later in the decade.

Historically, it marked a first in western technical cooperation. It seemed to project a NATO in space. Although no military significance was attached to the Spacelab-Shuttle partnership, it created a Western alignment in space operations vis-a-vis the space consortium the Soviet Union was building with its East European vassals.

Since the Apollo-Soyuz Test Project of 1975, when American and Russian spacecraft linked in an orbital exhibition of detente, spacefaring by East and West had become increasingly polarized, in step with their relations on the ground. Their disengagement increased the likelihood of space becoming an arena of military operations, of space war, an unthinkable eventuality that every military establishment had to think about.

Although the ESA-NASA alliance on Spacelab appeared to be a benchmark in cooperative space effort, it did not rule out competition. France and other European interests were not content to put all their space eggs in the Shuttle cargo bay. Nor were they content to depend on the United States for launch service.

There was a strong feeling in ESA that despite the failure of the defunct ELDO to develop the Europa rocket in the 1960s, Western Europe should take another shot at building its own launch vehicle for its own satellites.

Consequently, the ESA nations agreed in 1975 to put up 3.4 billion francs, or $714 million, to develop a heavy launch vehicle of their own. It was a three-stage rocket with the cosmetic name of Ariane. Its primary purpose was to free Western Europe of dependence on the Shuttle or expendable U.S. launchers for the placement of European commercial or scientific satellites.

The launch site ESA would use was the one France had built at Kourou, French Guiana, in the 1960s for the Europa rocket.

The European Space Agency's Ariane launch vehicle was designed to compete with the Shuttle for satellite payloads.

(ESA)

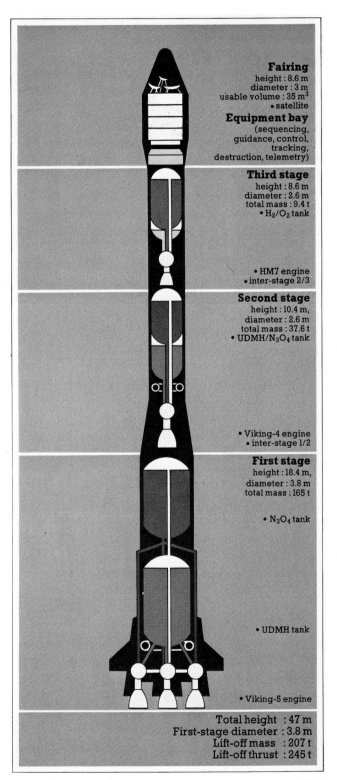

Fairing
height : 8.6 m
diameter : 3 m
usable volume : 35 m^3
• satellite

Equipment bay
(sequencing,
guidance, control,
tracking,
destruction, telemetry)

Third stage
height : 8.6 m
diameter : 2.6 m
total mass : 9.4 t
• H$_2$/O$_2$ tank

• HM7 engine
• inter-stage 2/3

Second stage
height : 10.4 m,
diameter : 2.6 m
total mass : 37.6 t
• UDMH/N$_2$O$_4$ tank

• Viking-4 engine
• inter-stage 1/2

First stage
height : 18.4 m,
diameter : 3.8 m
total mass : 165 t

• N$_2$O$_4$ tank

• UDMH tank

• Viking-5 engine

Total height : 47 m
First-stage diameter : 3.8 m
Lift-off mass : 207 t
Lift-off thrust : 245 t

Ariane diagrammed.   (ESA)

Once Ariane was operational, ESA believed it could launch satellites into geosynchronous orbit from Kourou at less than NASA's Shuttle charge. The ESA rocket program thus threatened to compete for payloads with the Shuttle.

By the late 1970s, the ESA-NASA relationship was cooperative on the one hand and competitive on the other. In late 1977, ESA offered the International Telecommunications Satellite Organization, an association of 102 nations using the machines, a cut rate for launching the big Intelsat 5 satellite with Ariane in 1982. The rate ESA offered was $1 million less than that estimated by NASA for a Shuttle launch. The deal was highly speculative, for Ariane was two years away from its first test launch.

ERNO

Two major projects, Spacelab and Ariane, dominated ESA's space activity. Ariane commanded the higher investment of the two. It promised to make ESA a spacefaring power.

In proportion to its investment of 53.3 percent, West Germany had the lion's share of the contracts to build Spacelab. France had the major share of Ariane development, which was managed by the French national space agency, Centre Nationale d'Études Spatiales (CNES).

The prime contractor for Spacelab was ERNO, the space division of VFW-Fokker, an amalgam of German and Dutch firms. The ERNO plant is on the edge of Bremen, across the street from the airport. It employed about a thousand engineers and technicians to construct and integrate Spacelab. About a half mile away was the burned-out shell of the Fokker-Wulf airplane factory, which was bombed by the British in 1944 and never rebuilt.

ERNO was organized from a cluster of North German aircraft companies, principally Focke-Wulf, Westerflug, and Hamburger Flugzenbau. They merged into a working group called Entwicklungspring Nord in order to take part in ELDO's Europa rocket development.

Since 1970, ERNO has been a fully affiliated company of the Zentralgesellschaft VFW-Fokker organization of Düsseldorf, which absorbed the older VFW group. VFW-Fokker has Dutch and American affiliations. It manufactures North Sea oil rigs, submersible vessels, and commercial aircraft, as well as rocket engines (the second stage of Ariane) and Spacelab.

The Number 5 Streetcar

It was raining, as it did almost every day in May 1978 in the Bremen-Hamburg area, as I steered a rental Ford Fiesta across the Weser River bridge into downtown Bremen. Traffic was edging along bumper to bumper. It was late afternoon, rush hour, I suppose. The rain made for poor visibility, and the streets gleamed with headlight reflections, but the street lights had not yet been turned on. Street signs were invisible in the murky light. We were swept along with the traffic stream, hoping eventually to see our hotel. Luckily, the hotel had an illuminated sign and an arrow pointing to an alley that led to a parking lot in the rear. There was one place left, just big enough for a little car.

Engineers at the Kennedy Space Center in Florida who visited Bremen to consult with ERNO and ESA engineers about the interfaces between Spacelab and the Shuttle had advised that the easy way to reach ERNO was to take the Number 5 streetcar.

It was still raining early the next morning as I joined a queue in front of the railroad station to await the streetcar. The hotel porter had assured me it would arrive at 8:11 A.M., and it did, by the station clock.

There was no need to peer out of the fogged-up windows for landmarks, because ERNO was at the end of the line. By the time it got there, the car was filled with umbrella-armed workers reading the *Weser Kurier*.

The streetcar itself exhibited a curious blend of the antique and the modern that links European cities with a medieval past that is missing in America. It was an old-fashioned trolley car of the vintage of those that used to ply Euclid Avenue in Cleveland 30 years earlier, yet there was nothing antiquated about its operation. It ran as smoothly as though it had just come out of the factory.

All out! ERNO. This firm was one of two West German competitors for the Spacelab prime contract. The other was the south German industrial group known as MBB, for Messerschmitt-Bölkow-Blohm. At ESA headquarters in Paris, I had been told that although MBB rated higher than ERNO in design and technical ability, ERNO was judged to have the edge in management. The factor that weighted the ESA decision in ERNO's favor most heavily, however, was the relatively high unemployment of skilled workers in North Germany compared with relatively full employment in the Munich industrial area. It would be easier to recruit a labor force in Bremen under these conditions than in Munich. ERNO was awarded the prime contract on September 30, 1975, for 600 million marks ($206.25 million) on a cost plus basis.*

Three American aerospace firms which had prepared early designs of the RAM project—McDonnell-Douglas, TRW, and General Dynamics—were engaged by ERNO as consultants.

Clean Room

Visiting a European aerospace plant requires the same formalities as visiting one in the United States. When properly identified, the visitor receives a badge that usually clips onto the breast pocket. In the 20 years that I have been writing about space programs and touring government centers and industrial plants, I have always received a badge. This is an important ritual, for the badge possesses a kind of "mana," or elemental force, in our industrial society; it is an emblem that confers upon the wearer a certain status, surrendered only when one returns the badge on leaving. I have a whole drawerful of badges I have walked away with.

During the morning, my guide was an English engineer, Peter Colson, who had worked on the Concorde and now represented ESTEC at ERNO. An engineering model of Spacelab rested in a steel scaffold near the center of the Integration Hall, a vast, brightly lighted chamber that echoed like a gymnasium.

We paused in the vestibule to don white nylon smocks

*ERNO and Messerschmidt-Bölkow-Blohm merged in 1983 and the Spacelab prime contractor became MBB/ERNO.

and military-style overseas caps of white paper. The Integration Hall is a huge "clean room" which is kept as dust free as possible. An air blower dusted off our shoes before we could step into it. Clean rooms are a standard environment in space work.

The engineering model consisted of two cylinders in tandem, open at each end, made of polished aluminum. Each was bigger around than it was long. They formed the outer shell of the Spacelab workshop module.

Each cylinder was a segment of the workshop. It was 8.85 feet long and 13.31 feet in diameter, giving it the shape of a drum. Two segments joined end to end formed the "long module" of the workshop, making it 23 feet long, including the end cones. One segment contained a "core" of subsystems for electrical power distribution and environmental control. It also provided a limited volume of laboratory working space and could be flown alone as a "short module." The second segment would contain only experiment racks and equipment. The racks were c-shaped so that they could be "loaded" with equipment and supplies outside the module and then slid into it before flight. This would be done at the Kennedy Space Center.

Integration hall of VFW-Fokker/ERNO at Bremen, German Federal Republic, where Spacelab was assembled. At the center is a cylindrical module segment. Two of these segments make up the long module of the pressurized laboratory. An equipment rack that slides into the cylinder is shown at right. (ERNO)

Technicians inspect the long module of Spacelab in the integration hall of VFW-Fokker/ERNO, Bremen.   (ERNO)

I followed Colson up a shaky aluminum ladder of the workshop, or "raumsfahrtlabor" (space travel laboratory). Two electricians were lying prone on the metal floor, installing wire bundles. They ignored us as we stepped over them.

Standing inside a segment without the experiment racks made it seem very large. There was plenty of room in two segments for three specialists to work in orbit without crowding. During flight, three persons in the long module would be the maximum for an eight-hour shift. The limitation was imposed by the heat rejection capability of the system. Human body heat is a factor to be reckoned with in any space vehicle.

One of the mystifying aspects of space work to outsiders is the paucity of workers it seems to require at any one time. We are accustomed to think of industrial production in terms of assembly lines and much noisy activity. By contrast, the "clean room" is a tomb. Only a few technicians ever seem to be working there at any one time. Often, they are to be seen conferring with one another or on the telephone or mulling over sheaves of blueprints and other documents.

Counting Colson and me, there were six persons in the Integration Hall that morning. Was it a holiday? No. Colson assured me that the work is carefully scheduled, so precisely in sequence that only a few technicians are required to perform at any one time.

Much of the work had already been done by subcontractors. The shining cylinders had been rolled, forged, shaped, and welded by Italian craftsmen at the Aeritalia aircraft plant at Turin and shipped to Bremen across the Alps by truck. Italy, with an 18 percent investment in Spacelab, was the major subcontracting state for the pressurized module at a price of 94 million marks ($46 million). The two cylindrical segments on the scaffold would contain a sea-level atmosphere at 14.7 pounds per square inch of pressure.

The habitability segment of the long module was equipped with an airlock. It would enable crew members or experimenters to expose certain experiments and instruments to space, as well as to climb out of the mod-

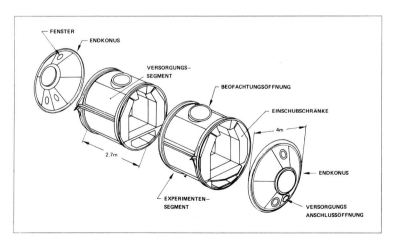

Modular structure of the Spacelab pressurized module, with two segments and end cones.   (ERNO)

Artist's portrayal of scientists working in Spacelab pressurized module. Cutaway shows instrument-experiment rack and tunnel leading to the orbiter cabin.
                                                                                            (TRW, Inc.)

ule for extravehicular activity (EVA) in the Shuttle cargo bay.

A complete mock-up of the workshop long module, with racks installed, had been erected at one end of the Integration Hall. With the experiment racks in the walls, the workshop looked like the interior of a railway mail car. The experimenters could sit or stand—not much difference in microgravity (so-called zero gravity)—in a wide corridor between the racks. In the long module, there

was enough aisle room for a "zero-gravity sled" to run on a track the length of the module. An experimenter would sit on the sled and be subjected to its rapid back-and-forth movement as a test for susceptibility to space adaptation syndrome (motion sickness) in orbital flight. The syndrome, with symptoms of nausea and vomiting, has been linked to vestibular (inner ear) disturbance. It has caused more loss of working time on manned space-flights than any other physiological problem. One of the

Three European payload specialists selected in 1977 by the European Space Agency stand in a full-scale model of the Spacelab module. They are (left to right) Claude Nicollier, Switzerland; Ulf Merbold, West Germany; and Wubbo Ockels, Netherlands. Merbold would be the ESA payload specialist on the first Spacelab mission. (ESA)

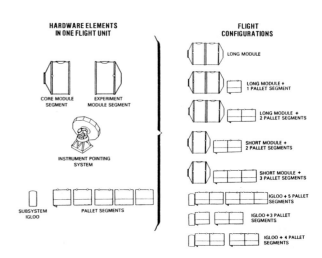

Spacelab segment configurations can be varied for specific Shuttle missions. The first Spacelab mission was to consist of the two-segment long module and one pallet segment. The igloo, a pressurized cylinder, contains subsystems necessary to operate pallet experiments when only pallets are flown.

(ERNO)

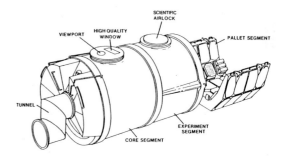

Drawing of the Spacelab long module and pallet configuration, selected for the first mission. (ESA)

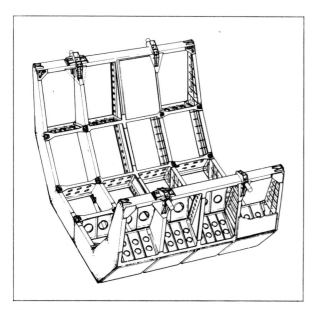

Pallet structure, made by Hawker Siddeley Dynamics, United Kingdom. (Hawker Siddeley)

medical experiments planned for Spacelab was to explore treatment for it by exposing volunteers to accelerations on the sled.

Along a wall of the Integration Hall stood an array of U-shaped aluminum platforms, each 9.5 feet long. They were instrument pallets, which could be carried with the workshop module or separately in the cargo bay. An array of telescopes, radiometers, cameras, and sensors could be mounted on the pallets and monitored from the workshop, from the orbiter's cabin, or from the ground.

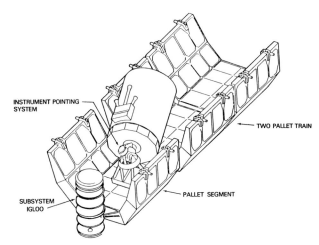

A two-pallet train with the instrument pointing system and igloo subsystems.   (ESA)

The workshop segments and pallets could be arranged in the orbiter's 60-foot cargo bay in a number of ways. On Spacelab 1, the long workshop module and one pallet were to be flown. Spacelab 2 was planned to be a pallet-only mission, devoted to astronomy and astrophysics.

Colson and I were accompanied to the plant lunchroom by two German engineers and Achim Nordmann, ERNO's ambassadorial public relations man. Nordmann exhibited a deep sense of awe about the project. "Here you see one of the brightest parts of the future being prepared in the Federal Republic of Germany," he said, as we walked between the factory buildings in a light drizzle. The work at ERNO involved great expectations for Western Europe, Nordmann assured me, and these could be realized only with German participation. ERNO was not simply a link to these great events, but a prime mover, I was told. Local pride came to the surface in the disquisition. "In this part of Germany," Nordmann said in a stentorian declamation, "in north Germany, the way we do things is the right way."

As prime contractor, ERNO's role was the integration (assembly) of parts made in ten countries and the subsequent testing of 13 major subsystems. The subsystems were manufactured following the completion of the critical design review in 1977.

Aeritalia fabricated the module structure of aluminum waffle sheet metal, which was milled, molded, and welded in the Turin plant. The pallets were manufactured in England by Hawker-Siddeley Dynamics under a 50-million-mark ($17.2 million) subcontract. The airlock in the habitability segment of the workshop was manufactured by the Dutch plant of VFW-Fokker under an 18.8-million-mark ($5.4 million) contract.

A Spanish firm, Sener, developed mechanical ground support equipment under a contract for 30 million marks ($10.3 million). ERNO made the habitability segment subsystem, which includes handrails, footholds, dust containers, workbenches, tools, and plating.

Two thermal control systems were installed by Aeritalia under a 75-million-mark ($25.8 million) subcontract. One consisted of cooling equipment for the experiments; the other is a "passive" system of paint and insulation to ward off sunshine.

The life-support system subcontract was awarded to Dornier of Friedrichshafen for 130 million marks ($44.8 million). The system is designed to circulate air and regulate temperature between 18° and 27° C. (64° to 80° F.), remove odors and pollution from the workshop atmosphere, and operate a fire warning and extinguishing system. Life support was designed for two persons continuously or for three for 8- to 12-hour shifts.

Heat generated by electrical equipment and crew members during flight would be disposed of by radiators on the inside of the Shuttle's cargo bay doors, which dissipate heat for the orbiter cabin as well as for the workshop.

Information from the experiments will be collected, displayed, and stored by command and data subsystems built into the workshop module. A 180-million-mark ($62 million) subcontract for this equipment was given to MATRA, a French electronics firm. Parts of the system were subcontracted out by MATRA to other firms in France and Germany. One American company, ODET-ICS, received the subcontract for a mass memory unit. ERNO's rival, MBB, was engaged as subcontractor for a high-rate multiplexer. The subcontract for the main computer in the system was awarded to CIMSA a French firm.

AEG-Telefunken, the German electronics company, received the subcontract for electrical power distribution in the core segment of the workshop. The 60-million-mark ($23 million) contract called for a 28-volt direct current system drawing power from the orbiter's fuel cell batteries.

Electrical ground support equipment was manufactured by the Bell Telephone Company of Belgium under a 70-million-mark ($24.1 million) subcontract. Bell subcontracted part of the work to Compagnie Industrielle Radioelectronique, Switzerland; Études Techniques et Construction Aerospatiale, Belgium; and two Paris firms, CIMSA Schlumberger and P. Fontaine.

Finally, fabrication of remote control and monitoring systems for the workshop and a stovepipe-shaped tunnel connecting the workshop with the orbiter cabin was undertaken by ERNO itself.

Along with Ariane, Spacelab became the matrix for the development of West European space technology, which hitherto had been confined to satellites, except for the defunct Europa rocket project. With Spacelab, Europe had a manned space vehicle and with Ariane, a technical start toward creating a launcher that could put it in orbit. The possibility of an independent European manned space program came into existence for the first time.

ESTEC

The principle of a strong central management was promptly imposed on Spacelab's international set of subcontractors. Although they were no more widely dispersed geographically than American subcontractors on the Shuttle, ESA had to deal with language barriers and a lack of standardized tools and methods. Could a Flemish electrical contractor perform in the same way as an Italian firm? It would if the big boss at headquarters said it had to do so.

A Tower of Babel effect threatened the efficiency of multinational space efforts in Europe. English was adopted as the official language in ESA for engineering development, while French served as the main administrative language, especially at ESA headquarters in Paris.

Management headquarters was set up at the European Space Research and Technology Center (ESTEC) in the Netherlands on the North Sea coast near the town of Noordwijk. The modern residential community rises discreetly from a flat landscape with tulips (in season); a big old windmill seems to have been left standing for tourists.

The multiwing, four-story structure that houses ESTEC has been built behind a mountainous sand dune that keeps out the North Sea. The central core of the building, about two city blocks long, contains administrative and drafting offices. There is an elegant cafeteria-dining room on the top floor. It is the only government-operated lunchroom I have visited where you can add fine wines and champagne to your tray. Housed in the wings that radiate from the central core are laboratories and shops. Satellites in various stages of development were visible behind glass panels, like caged birds.

It was not difficult to understand why ESA wanted its own launcher capability. Its scientific and utility satellites were highly sophisticated. The effort to develop a European launcher in the 1960s had failed largely for the lack of a strong central management. That mistake would not be repeated.

The Spacelab boss at ESTEC during the later part of the construction period was Bob Pfeiffer, a fullback-sized German engineer who seemed to be able to resolve difficult problems by plunging straight through them. Here was a boss who was fully aware that the buck stopped with him. He left no doubt who was in charge. He was.

There was no question that reusability was a major design goal of Spacelab. Wolfgang Nellesen, one of the lead engineers on the program, emphasized that a firm requirement for the design and manufacture of the workshop modules and the pallet was durability. Each unit was to have a lifetime of 50 missions, or ten years.

ESA had taken the Marshall concept of RAM at face value and was trying to develop it in every particular. It would provide scientific and industrial experimenters with a versatile laboratory and observatory that might last for half a century and was flexible enough to be converted into a free flyer.

Spacelab could also be developed into a multiple-

module space station. But its initial purpose was to demonstrate whether the Shuttle-Spacelab combination gave the spacefaring countries of the Western Alliance a tool with which to develop the commercial and industrial potential of space near Earth.

The Scientific Questions

Ostensibly, the NASA-ESA partnership was a long step toward international technological cooperation which was essential for extended manned space exploration. Its only comparable precedent was the Anglo-French Concorde joint venture. The economic potential of exploiting the interplanetary medium was a common interest among the industrial nations of the North Atlantic. The development of Ariane, however, indicated the onset of competition between ESA and NASA for payloads.

The Marshall Space Flight Center was given the responsibility of seeing that Spacelab was built so that it could be loaded neatly into the orbiter's 60-foot cargo bay. A Marshall engineering team was posted at ESTEC to monitor the interfacing of the workshop module and pallets with Shuttle systems. There was a direct-dial telephone connection between ESTEC in Noordwijk and Marshall in Huntsville, Alabama.

Spacelab 1 initially was scheduled to fly in the latter half of 1980, but the flight date kept slipping with delays in the Shuttle. The long module and pallet would occupy a little more than half of the length of the cargo bay and would be linked to the orbiter cabin by a stovepipe-like, pressurized tunnel through which crew specialists could float when they changed shifts.

A total of 42 experiments had been planned in 1978, of which 25 were ESA's and 17 NASA's. One American and one European scientist would go aloft in the orbiter to work in the laboratory as payload specialists and one American and one European payload specialist would monitor experiments on the ground.

In 1979, the number of experiments was cut to 38 because of weight and data transmission limitations. ESA had 25 and NASA 13. Spacelab 1 itself was an experiment. It would demonstrate the utility of a captive laboratory, one tied to the Shuttle, in low Earth orbit.

The experiments addressed three scientific and technical areas: astronomy and solar physics; the physics of space near the Earth and of the upper atmosphere; and the effects of microgravity on living organisms, including man, on biochemical reactions, and on manufacturing processes. Similar experiments had been carried out on the Apollo and Soyuz spacecraft and Skylab and Salyut space stations. The results looked promising. Although some of the astronomy and space physics experiments duplicated those on satellites, Spacelab would bring the instruments and data recordings back. As Skylab had demonstrated, the return to Earth of data, especially photography, was a great advantage.

One group of experiments was of particular interest to weather and climate scientists. American, French, and Belgian investigators sought to determine whether there is any change in the intensity of sunlight reaching the top of the atmosphere. This is called the *solar constant* because of the long-held belief that it does not vary from a value of 1.94 calories per square centimeter per minute. However, variation in this value was observed in data from sounding rockets from June 1976 to November 1978. Some variation was also detected in data from the *Solar Maximum Mission (SMM)* satellite which was launched from Cape Canaveral early in 1980.

The solar constant would be observed by three instruments on the Spacelab pallet: An active cavity radiometer provided by NASA's Jet Propulsion Laboratory, which was designed to measure total solar irradiance and any variation in it; a triad of ultraviolet, visible, and infrared light detectors provided by the Service d'Aéronomie du Centre National de la Recherche Scientifique (France) and mounted on the pallet to determine which wavelengths of sunlight exhibited variation; and a high-resolution pyrheliometer (sunshine meter) provided by the Institut Royal Météorologique de Belgique (Belgium), which was designed to measure the solar constant directly from the pallet.

The reusability of Spacelab-Shuttle would enable scientists to continue these observations from time to time over periods of months, years, decades. Does the solar constant vary, and if so, does it vary enough to have been a major factor in long-term climatic changes, such as ice ages and interglacial periods?

From Spacelab to Space Station

On both sides of the Atlantic, Spacelab was viewed as a step toward a long-term or permanent space station. Initially, its tour would be 10 days. Both NASA and ESA planned to enhance life-support and power systems so that missions could last 30 days.

During the spring of 1978, I discussed the future of Spacelab with Heinz Stoewer, a German engineer at ES-TEC in charge of future development. It was a cool, cloudy morning in May, and the last tulips were fading in the little rectangular fields of Holland. At the ESTEC main gate, the guard remembered me from a visit the year before as I signed in and received the inevitable badge. Not many American reporters had visited the great research center.

Stoewer had worked for years in the American aerospace industry and was bilingual in English and German, an asset in Western Europe, where much high technology dialogue is conducted in both languages. A tall, rangy engineer, he would be indistinguishable in a crowd of Texans at the Johnson Space Center in Houston. Stoewer had drafted a "Spacelab Follow-On Development Program," which considered two options for additional power supplies. One was a small solar-powered satellite, compact enough to be carried into orbit by the Shuttle, where it would be deployed as a free flyer.

With its solar wings continuously converting sunlight into electricity, this "power module" was designed by the Marshall Space Flight Center to produce 25 kilowatts of direct current, 14 for the orbiter and 11 for Spacelab. Parked in orbit, it would be available for many missions. In theory, it could last for years, decades—no one knew how long. Space is very kind to equipment. Nothing rusts or corrodes. Early communications satellites passing through the Van Allen radiation zone sustained electronic damage from high-energy particles. This seemed to be the main threat to the longevity of a power module in low Earth orbit, aside from cosmic ray particle impacts and space dust erosion.

A second option, explored by the Johnson Space Center, was the solar cell wing. It consisted of a folded solar cell array, or wing, that would be lifted out of the orbiter cargo bay by the remotely-controlled manipulator arm

This solar array wing, 105 feet long and 13.5 feet wide, was manufactured for the Marshall Space Flight Center by the Lockheed Missiles & Space Co. as a means of providing electric power for extended Spacelab missions and for solar electric propulsion units.   (NASA)

and extended. When deployed, the wing would be 105 feet long and 13.5 feet wide. It would have 82 panels of photovoltaic cells. It would remain attached to the orbiter. The wing would be designed to produce 12.5 kilowatts. For the return to Earth, it would be folded back into the cargo bay.

The first step toward developing the wing was taken by NASA in 1978. It awarded a $2.7 million contract to the Lockheed Missiles & Space Company for the construction and delivery of a flight experimental wing by May 1980.

ESTEC's long-range view of Spacelab assumed a long period of cooperation with NASA. ESA also expected to sell Spacelabs to users other than NASA, which had agreed to buy at least one in the 1973 memorandum of understanding that formalized the partnership.

Stoewer's program visualized Spacelab as becoming a free flyer. Plugged into the power module, the workshop could remain in orbit 30 days or longer without the Shuttle. It would require living accommodations, which might be provided by an additional segment attached to the habitability module airlock.

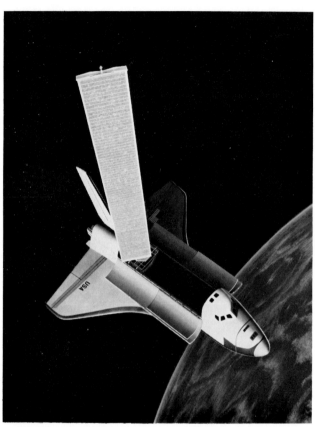

An artist's portrayal shows how the solar array wing would be deployed from the orbiter cargo bay. Tests of the wing have been planned for the mid- to late 1980s.   (NASA)

Spacelab need not be tenanted all the time in orbit. It could be parked there after a mission and revisited by teams of experimenters, as Skylab had been.

The final evolutionary step was not a big one. Stoewer pointed out that the Spacelab pressure shell was "a good candidate" for a cluster of space station modules that could be launched into orbit by the Shuttle. As many as five segments, with a total length of about 45 feet, could be carried in the orbiter cargo bay at one time. They could be joined in orbit to become the nucleus of a multimodule station that could be enlarged with the attachment of additional segments.

Another use for the workshop module could be visualized. Two or more segments could provide crew ac-

commodations on interplanetary voyages. Stoewer's report did not go that far, but a similar option utilizing space station modules was cited by the Space Task Group Report. In the Spacelab configuration, it was quite viable. Also, it would spread the cost of an interplanetary mission across the Atlantic. With Spacelab, a start had been made.

The Spirit of '75

With the Shuttle and a sort of space station in one package, NASA could once more afford the luxury of projecting future possibilities. It called on its centers for ideas: Ames Research Center and the Jet Propulsion Laboratory in California, the Johnson Space Center in Texas, the Lewis Research Center in Ohio, the Marshall Space Flight Center in Alabama, the Goddard Center in Maryland, and the Langley Research Center in Virginia. All were populated with scientists and engineers who could think far out. It was the far out that headquarters in Washington seemed to be looking for.

The contributions from the centers resulted in a thick book of proposals called *Outlook for Space* (1975), which the Government Printing Office issued in a dull brown cover. I summarize it here because it is the only program NASA has devised since the Space Task Group Report of 1969.

With Shuttle and Spacelab, *Outlook* suggested it would be possible to attack the basic questions of natural science: the origin and evolution of the universe, of the Earth, of the Moon, and of the planets. If these objectives sound familiar to space buffs, it was only because they had been cited a decade earlier as part of the scientific rationale for exploring the Moon in the Apollo program.

The study strongly suggested that the geocentric focus of mankind inevitably would shift to a heliocentric one. Man would consider the Solar System his domain, not merely the Earth. To this end, the development of a space transportation system was an essential feature of human development.

An analysis of the world energy crunch clearly pointed to the necessity of harnessing solar power in space, by means of satellite power collectors beaming electrical energy to the ground in the form of microwaves. But this kind of power source requires a much bigger investment in transportation than was represented by the Shuttle. Eventually, it will have to be made if mankind is to survive even another millennium in a civilized state, unless fusion power on a large scale is achieved. The distinction between solar and fusion power is feasibility. Solar energy collection and distribution are within the state of the art; fusion lies beyond it, perhaps by a century or more.

The predominant opinion among ecologists and fuel technologists is that the age of fossil fuel is ending. A number of surveys contend that oil and gas supplies will dwindle beyond the range of economic feasibility in another half century. Coal would provide energy for 500 to 600 years, but its pollution and carbon dioxide emission would degrade the planetary atmosphere and alter the climate.

Fissile fuels—uranium, thorium, plutonium—appear to be a short-term solution to mankind's longer-term energy requirements. Even with breeder reactor technology, which would stretch uranium resources fourfold, the ultimate supply is limited. The potential health hazard of the breeder reactor, which converts uranium 238 into fissile plutonium 239, a deadly carcinogen, imposes risks that many find unacceptable.

The immediate "fix" for the energy gap appears to be solar power. Because of the day-night cycle, clouds, dust, and air pollution, the use of ground-based solar collectors on any big scale appears economically questionable. The more efficient way is to establish sunlight collectors in geosynchronous orbit, 22,300 miles over the equator.

A collector would consist of an array of photovoltaic cells, which convert direct sunlight into direct electrical current. The electricity would then be converted to microwave energy and beamed to Earth by a transmitting antenna. The microwaves would be received by antennas on the ground and reconverted into electricity for distribution through conventional systems.

A number of engineering studies of solar power satellites have been made by NASA, its contractors, and con-

sulting firms. Department of Energy funding for "SPS" research and development has been disappointing, largely because of the lack of priority for such a program.

How a solar power satellite would function and how it could be built are described in chapter 8. The essential point is that any system for obtaining power from space on a large scale requires a Shuttle-based space transportation system.

Without the Shuttle, and auxiliary vehicles to come later, we are locked into the resources of our own planet. And these, even with ground-based solar energy collection systems, may not be sufficient to meet the energy requirements of civilization in future centuries.

A Moon Colony

Outlook for Space suggested that large space structures, such as solar power satellites, might be fabricated on the Moon at considerable savings in transportation cost. Moving materiel into geosynchronous orbit from the Moon would be 20 times cheaper than boosting them off the Earth, because of the Moon's lower gravity.

Power is available on the Moon from thorium, which can be converted into uranium 233, or from an array of photovoltaic cells laid out on the airless lunar surface. The solar cells could be made from lunar silicon. Oxygen is available from the rocks, but hydrogen is scarce and would have to be imported unless ice is discovered in shadowed regions of the lunar poles.

Beyond the Moon, a Shuttle-based transportation system might be sufficiently developed in the next century to make it possible for prospectors to look for iron, nickel, and other metals in the asteroid belt. An iron-nickel asteroid just one kilometer in diameter would contain about four billion tons of high-quality iron, worth more than $400 billion on the 1975 market, *Outlook* suggested. If it could be moved by powerful rocket boosters into an Earth orbit, it would provide a source of steel for centuries.

Outlook considered human colonization of Mars and Venus as well as of the Moon. Mars is cold and dry and Venus is hot and dry, but *Outlook* set forth the hypothesis that these planets could be *terraformed*—that is, made Earthlike.

On Mars, the thin atmosphere might be increased in density by melting the large polar ice caps with nuclear or solar heaters. Raising the density of the Martian atmosphere would raise the average temperature and make it possible for free water to exist on the surface. There is abundant evidence that water flowed over the surface of Mars in the far past, when the atmosphere must have been considerably denser than it is now. In fact, there appear to have been bodies of water on Mars the size of the Caspian Sea. Runoff channels as large as earth river valleys apparently were formed by running water in the equatorial uplands. The evidence for water implies a period in the past of a warmer climate.

Several investigators have noted evidence of catastrophic flooding. It appears to have discharged water through large channels at the rate of a billion cubic meters a second—ten thousand times the mean annual discharge of the Amazon River.[1]

In time, plants could be grown to reduce the carbon dioxide atmosphere and add oxygen to make it more Earthlike.

Venus, with its massive carbon dioxide atmosphere 100 times denser than Earth's and its surface temperature of 1,000° F., presents a more formidable problem.

The study cited a theory that much of the carbon dioxide could be absorbed by seeding the surface with algae from Earth. The terraforming of Venus seems to be a challenge for future centuries.

Beyond the Solar System, the study considered the prospect of interstellar flight. Present propulsion systems are not capable of reaching the nearest star, Proxima Centauri, four light-years away. The study conjured up an antimatter engine, gaining energy from the annihilation of matter by contact with antimatter. The total conversion of matter to energy would provide an interstellar propulsion system. The idea has been around for years, in science fiction. But applying it is somewhere beyond the twentieth or twenty-first century.

The little brown-covered book tells us what a number of thoughtful people in the NASA centers are thinking about. But at the end of 1975, the reality was a transport limited to low Earth orbit and a Spacelab limited to seven- to ten-day space tours.

Moreover, the engine contractor, Rocketdyne, had become bogged down in engine testing failures. Before we could go anywhere in space, the Space Shuttle Main Engine had to be developed to a confident level of performance. It wasn't working, and that was giving the space agency and the contractor a very bad headache.

NASA's time of troubles had started.

From Success to Failing

Following the successful approach-and-landing tests in 1977, the *Enterprise* was hoisted atop its Boeing transport in California and flown to the Marshall Space Flight Center in north Alabama for seven months of structural testing.

Inasmuch as the orbiter was a product of aircraft technology, no surprises were expected. The tests consisted of vibrating the orbiter alone and also mated with its external propellant tank and solid rocket boosters to determine how well all parts of the Shuttle assembly would withstand flight vibratory stress. In general, although the tests showed that parts of the assembly required strengthening by adding more metal, they confirmed the engineering assessment that the Shuttle was structurally sound.

Together with the approach-and-landing experience, the structural tests confirmed the impression created by NASA in 1972 that developing the Shuttle would be a straightforward project in known technology; that it would

not require any dramatic breakthroughs; that it could be done largely with off-the-shelf parts. This impression contributed significantly to the difficulties, delays, and cost overruns that plagued the program in the later 1970s, because it allowed the space agency to underestimate the cost and the magnitude of the technical challenge of developing the first reusable space vehicle. And it persuaded Congress to go along.

During hearings on space transportation before the House Committee on Science and Astronautics in March 1971, Dale Myers, NASA associate administrator in charge of Shuttle development, had presented a gung-ho, optimistic view of Shuttle prospects. As of 1971, the schedule called for the first horizontal flight in 1976, the first manned orbital (test) flight in 1978, and an operational Shuttle in 1979, according to Myers' presentation to the committee.

Myers had projected the first horizontal flight with jet engines. After these were taken out of the design to save weight and cost, the approach-and-landing glide tests became the horizontal flight test equivalent. Coming a year later than the 1971 schedule had called for, they indicated that the orbital flight test program would be late also.

Warned that the Space Shuttle Main Engine (SSME) would require more "lead time" than other components, committee members asked Myers what difficulties the SSME presented. He assured them that previous Air Force and NASA experience provided the knowledge necessary to proceed with the design of the high-pressure hydrogen-oxygen engine, which would be the most powerful of its type ever built.

"The high-temperature materials required for use in this engine are already available," he said. "By beginning the design now, we can meet our planning schedules, which call for a flight-ready propulsion system to be available about two years prior to the operational readiness of the vehicle. Three definition studies of the main engine are in progress by contractors and will be completed in June 1971." The technical requirements "are in hand for the integrated electronics control system," he added, "but the applications of this technology will require innovation and proof of concept demonstration."[1]

Myers' testimony reflected the conviction of the NASA space transportation directorate. It was a belief rooted in the success of Apollo, the lunar module, and the Saturn 5 moon rocket. Any organization that could build that system could build any system, it was said.

The official optimism elicited a letter to Myers from Rep. Joseph E. Karth (D-Minn.), chairman of the House Subcommittee on Space Science and Applications. Karth asked about the technical challenges of Shuttle development: "During your appearance before the Science and Astronautics Committee, I received the impression that NASA officials feel confident that the technical problems associated with the development of the Space Shuttle are manageable and that the necessary technology is either in hand or within view of acquisition. Frankly, I found your testimony quite reassuring."

Myers replied, "I appreciate your comments concerning your reassurance that the technical problems asso-

ciated with the space shuttle are manageable and that solutions are within reach.

"That was certainly the impression and the message I was trying to convey because we in NASA do feel confident that although all the answers are not yet in hand, the progress indicates that they will be at the proper time to support a step-by-step development program."

The major technical challenges had been assessed in 1969, Myers said. They were the SSME, the avionics (computer-managed flight controls), and the thermal protection system, or heat shield. "No technology breakthroughs are required," Myers said. "Essentially, the Shuttle will be based on 1971–72 knowledge applied to allow us to fly an operational Shuttle in 1979. "Although significant innovation is required, principally to keep development costs low, the basis for the technology required is solidly in hand."[1]

The Good Vibes

Like the glide tests in California, the structural vibration tests in Alabama seemed to bear out Myers' forecast, even after that talented engineer left NASA to join the Department of Energy.

No other place in the Western world is as well equipped to conduct structural tests of a large vehicle as the Marshall Space Flight Center. It was built for NASA on the grounds of the U.S. Army's Redstone Arsenal in a rolling Appalachian valley near Huntsville, Alabama.

"Redstone" describes the soil of the region, which is as red as that of the planet Mars. It was the Redstone Arsenal that the Army selected for its rocket development program in the late 1940s. There most the German emigré engineers and technicians who had developed the V-2 for Hitler settled to build rockets for the United States under the direction of Major General John B. Medaris and their Peenemuende leader, Wernher von Braun.*

It is a remarkable turn of history that the German rocket team was able to utilize the high technology first of Ger-

* Peenemuende was the site of Nazi Germany's V-2 rocket development center on the island of Usedom in the Baltic Sea.

many and then of the United States to develop deep space rockets as well as missiles. The Saturn 5 booster stage, a product of Marshall, was a lineal descendant of the old World War II vengeance weapon, the V-2.

Marshall was also the site where Skylab, our super space station, was built, from the third stage of the Saturn 5. Medaris advocated in-house development by the government of high technology systems. "It keeps the contractors honest," he told me.

Because of these programs, Marshall evolved into NASA's principal static test center for big rockets, with a test annex on the Pearl River hard by the Mississippi Gulf coast near the town of Bay St. Louis. The annex evolved, too, to become the National Space Technology Laboratory (NSTL).

Both of these testing facilities were modified for the Shuttle. At Marshall, the giant test stands which had been built for the Saturn family of rockets were altered for the Shuttle, and at NSTL, the old Mississippi Test Center, the engine test stands were equipped for the SSME.

The main structural testing at Marshall was done in the Dynamic Test Stand, a steel and concrete building 415 feet high. It was a smaller version of the 550-foot-tall Vehicle Assembly Building (VAB) at the Kennedy Space Center in Florida. These giant structures are unmatched anywhere outside the United States and the Soviet Union. The VAB is one of the largest in the world, second in cubic feet to the Boeing plant in Seattle.

In their way, they reflect the power and purpose of a beginning spacefaring civilization, into which we seem to be slowly evolving, as the pyramids characterized the life-after-death outlook of classical Egyptian culture some four millennia ago. Our space structures are monuments to the future, however, instead of the past.

Enterprise, riding atop its 747 carrier, arrived March 21, 1978, at the airstrip serving the Arsenal and Marshall and was towed to the testing complex by a tractor along the divided highway that forms the main thoroughfare of the complex of test stands, hangars, laboratories, and the central headquarters office building. Hundreds of space workers turned out to watch it—the first spacecraft with wings.

During the late spring of 1978, *Enterprise,* mated with

its external tank, was suspended tail down in the 360-foot-high test tower for the Mated Vertical Ground Vibration Test. Push-pull forces were applied to spars, ribs, and the skin of the vehicle.

The degree of force was controlled by radio frequency amplifiers, analogous to those in a hi-fi. The first test phase began June 15 and ended July 14. It produced vibratory forces similar to those the orbiter and external tank would experience two minutes after liftoff, when the solid rocket boosters (SRBs) dropped off. The second test phase, which began October 6, added the SRBs to the stack. The entire shuttle was then put together, as it would be on the launch pad, for the first time. It weighed 4,193,200 pounds. The third test phase began in November. The fuel load of the

Forward assembly of a Shuttle solid rocket booster is lowered to be mated with booster elements at the Marshall Space Flight Center.   (NASA)

SRBs was varied to simulate the loading from liftoff to burnout about 120 seconds later.

I watched part of the first test at the end of June. The orbiter and tank were being shaken, but the vibratory effect was so rapid and small it could not be seen. It was measured by stethoscopelike pickups attached to the metal. The magnitude of the push-pull–induced vibrations was controlled and their effect was recorded on a bank of dials in a console in the ground floor control room.

During the lunch break, a brown bag–thermos bottle interlude, I squeezed into the test stand elevator to ride up to level 12, about 130 feet above the cement ground floor. From that level, clutching the shaky steel railing, I could look down on the nose of the *Enterprise* as it hung from the boom atop the stand. The sleek cylinder of the 154-foot-tall tank towered above me.

Without the SRBs, the *Enterprise* and its towering propellant tank looked like a broad-winged airplane riding piggyback on a large blimp. The assembly seemed awkward and primitive. It was a far cry from the double airplane system proposed in 1969.

To the observer, the stack seemed to be motionless, but actually it was vibrating at about 30 Hertz (cycles per second) in response to forces applied by exciter rods. These were designed to simulate aerodynamic forces the Shuttle would experience 120 seconds into the flight and at an altitude of 27 miles.

Sensors would detect transfer modes on the vehicle in which the vibrations would be amplified by resonance, the process by which vibrations beget vibrations of higher amplitude. One could think of it as echo. At some points in the structure, the echo was louder than the sound that produced it. These points identified areas that would be strengthened to damp out the echo, by adding more metal.

Eugene Cagle, the Marshall test director, told me that no structural problems had shown up. Minor modifications in the flight computer might be required to avoid vibratory interference in highly turbulent regions of the flight path.

Later, the tank was removed from the test stand. It was lowered by a huge crane and deposited in an eight-wheeled trailer painted bright yellow.

"We like to set an example as an equal opportunity employer," my guide commented, waving at the crane operator, a young woman who had taken off her hard hat to comb her hair.

The tank was the expendable component of the Shuttle. It would be dumped into the sea when its propellant was used up and the orbiter neared orbital altitude. In the early 1970s, it was estimated that when the Shuttle became fully operational, 60 of these huge tanks would be thrown away every year at the flight rate then projected for the late 1980s.*

Jack Connors, chief of the structures and propulsion test division, discussed some of the thinking at Marshall about the tank. It could be outfitted, as Skylab had been, as an orbital workshop. Living quarters could be partitioned off and there would be plenty of room for a workshop and even a hangar for small space vehicles. By linking several tanks, a large space station could be established in orbit. The aluminum alloy walls could be insulated with spray foam. A docking port could be placed on the liquid oxygen end of the tank. NASA headquarters did not buy the idea.

The ground vibration tests ended February 23, 1979. They resulted in structural modifications, such as strengthening brackets in the forward part of the solid rocket boosters to reduce vibratory effects on the guidance and gyrostabilizer systems. Although minor, the changes added weight.

Engine Trouble

Early confidence expressed in developing the Space Shuttle Main Engine gave way in 1976 to the grim realization that it was going to be one of the toughest jobs NASA and the contractor, Rocketdyne, had ever tackled. The Shuttle directorate underestimated the order of difficulty of developing this advanced, high-pressure engine, but once having started, it had to keep going. It had a technological bear by the tail. Engine troubles were to continue to delay and threaten the safety of the program right up to launch.

*By the end of 1981, the flight rate was reduced to 24 a year.

The SSME is the first rocket engine I know of that is designed to be reused 50 times or more, in an airline-style operation. All other liquid-fuel rocket engines and solid-fuel rocket motors are fired only once and then thrown away.

Reusability alone required a higher order of technology than had been achieved before in the space industry. In addition, the SSME would operate at higher pressure and for longer periods than any previous American main engine. The high-pressure fuel pump, the Rocketdyne people said, was capable of emptying an Olympic-size swimming pool in 25 seconds.

It was surely a "gee-whiz" engine, or would be if it could be made to work. Under the development schedule proposed in 1971, the SSME was to be a mature engine by 1977. Between March 24, 1977, and November 4, 1979, there were 14 engine test failures. They were caused by faulty seals, uneven bearing loads, cracked turbine blades, cracked fuel injector posts, heat exchanger malfunctions, valve breakdowns, and ruptures in the hydrogen lines. Eight of the failures resulted in fires which damaged the engines and, in some instances, the test stand and the mock-up of the orbiter rear fuselage in which the three-engine cluster (the main propulsion test article, or MPTA) was fired.

To NASA's dismay, the development of the SSME was recapitulating the test failure experience that has attended the building of every new rocket engine since the V-2.

Because NASA had not approached engine development with the expectation that it would require the test-and-fix process that characterized earlier rocket development, the agency seemed unprepared for the problems that had been considered routine in the past.

The SSME was viewed by the agency and the contractor as an advanced version of the J-2 hydrogen-oxygen engine that had been developed for the upper two stages of the Saturn 5 moon rocket. Consequently, NASA and Rocketdyne adopted a "success-oriented" strategy, which assumed that, because of previous experience with this type of engine, it would be economical to build it with a minimum of separate testing of the components, such as the seals, bearings, and turbo pumps.

Because the SSME represents a major technical advance in the space age, it is worth a rough description here. The technology so expensively acquired in making it work may soon be applied in the aircraft industry, which has considered a hydrogen-fueled power plant for a long time, especially in military planes.

The SSME

As designed for the orbiter, the SSME is the most efficient rocket engine known to the West. The next step up in efficiency is the nuclear rocket engine, which was abandoned in 1972, partly because of its potential radiation hazard and mainly because NASA had no plan to use it in this century.

The SSME is unique because it packs great power in a small package. Its thrust-to-weight ratio (specific impulse) is the highest of any chemical engine I know of. There are three of these hydrogen-oxygen engines in the tail of the orbiter. Each puts out 375,000 pounds of thrust at sea level (470,000 in vacuum), a total of 1,125,000 at liftoff—over one-sixth of the 6,495,000 pounds of thrust required to lift the Shuttle off the pad. The other 5,370,000 pounds is provided by the two solid rocket boosters.

Two minutes after liftoff, the SRBs burn out. They are dropped into the ocean by parachutes to be recovered and reused. The orbiter and its propellant tank keep on going under the thrust of the SSME propulsion system, which could increase to 1,410,000 pounds as the vehicle approaches orbital altitude. Then the engines shut down and the big propellant tank is dropped off; now we have a spacecraft.

The only existing hydrogen-oxygen engine of comparable size is the J-2. It generated 200,000 pounds of thrust for the upper stages of the Saturn 5 moon rocket. An uprated version, the J-2S, developed 250,000 pounds. The J-2S was considered for the orbiter, but the final design called for a more powerful engine. The SSME, putting out 1.5 times the thrust of the J-2S, seemed to be the answer. It was not merely an uprated J-2; it was a new engine and a new challenge.

There are two ways of increasing the lifting power of

a rocket engine. One is to make it bigger, but this increases the weight so that the thrust-to-weight ratio may remain the same or even decrease. The other way is to increase the pressure at which the gas molecules that provide the thrust are expelled from the engine nozzle.

This was the route NASA and Rocketdyne took. The power of the engine was dictated by the size and purpose of the orbiter. Its weight and payload called for the highest-pressure engine ever designed. It seems remarkable in hindsight that NASA adopted a success-oriented strategy for such an advanced engine, for such a strategy supposed that engine components would work the first time they were tested. Instead, Murphy's Law took over. It holds that in this type of program, anything that could fail would fail. A success-oriented manufacturing strategy was suitable, perhaps, for washing machines, refrigerators, and motorcycles, but it was certainly bad news for the Shuttle.

A New Steam Engine

In an engine burning liquid hydrogen with liquid oxygen, the combustion product is steam. Thus, the SSME is essentially a steam engine, although neither James Watt nor Robert Fulton might have identified it that way.

High-pressure steam generated by the rapidly burning fuel passes through a constriction chamber—the nozzle—and erupts from the throat at high velocity. Newton's third law does the rest: for every action there is an equal and opposite reaction. This is a steam reaction engine.

By raising the pressure at which the steam is expelled, the engineer gets higher thrust. But that is not a simple matter. This engine needed high-speed pumps to move the liquid hydrogen and liquid oxygen fast from the big propellant tank to the combustion chambers of the three rocket engines.

The turbine blades in the pumps turn at very high speed. They are probably the most advanced rotating machinery in the aerospace industry. Rotating machinery has invariably produced problems in reaction engines, from jets to rockets. The faster it runs, the more likely it is to break down.

The SSME was designed to boost combustion chamber pressure to 3,000 pounds per square inch (psi), 1½ times the pressure in the J-2. It was a big jump, and it brought big problems.

Propellant flow pressure was built up by two sets of pumps, each set consisting of a low-pressure and a high-pressure unit. The low-pressure pump raised fluid pressure to several hundred pounds per square inch and delivered the fluid to the high-pressure pump, which raised the pressure to thousands of pounds per square inch.

In the hydrogen pumping set, pressure peaked at 6,200 psi in the high-pressure pump. In the oxygen set, it was boosted to 4,600 psi. The turbine blades, rotated by the force of a mixture of hot hydrogen and superheated steam, ran under enormous stress. During early tests they developed metal-fatigue cracks.

Although rocket turbopump machinery was believed to be highly developed in this country, no pump had ever been built to operate at the extreme pressures of the SSME. Special materials had to be found; pumps had to be redesigned after they kept breaking down; assemblies had to be strengthened; bearings had to be improved.

Prior to 1970, the need for a high-pressure hydrogen-oxygen engine had been foreseen by NASA, which awarded a contract to develop one to the Pratt & Whitney Division of United Technologies, the Air Force Rocket Propulsion Laboratory, and the Marshall Space Flight Center.

A prototype high-pressure experimental engine was designed by this group, based on research in high-speed turbine machinery by Pratt & Whitney and Marshall. The engine was designed to produce 350,000 pounds of thrust. It was considered for the orbiter. However, NASA selected Rocketdyne in 1971 to develop the engine, before the Shuttle was authorized. It was explained that the engine would require more time to develop than other systems. Presumably, the research experience on the Pratt & Whitney engine was available to Rocketdyne through Marshall. Why the space agency selected its engine contractor in this way is not clear either to me or to other observers. Pratt & Whitney was not considered as a bidder. In fact, NASA later negotiated the engine contract with Rocketdyne without going through the bidding formality, mainly because Rocketdyne had built the J-2.

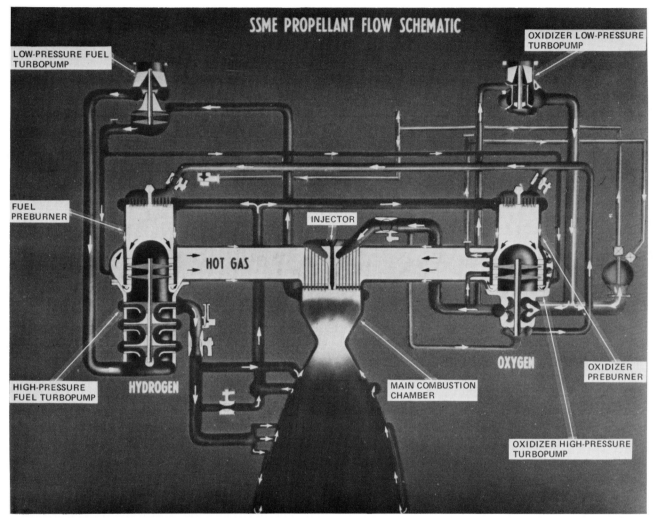

Diagram of the Space Shuttle Main Engine.   (Rocketdyne)

At the outset, NASA and Rocketdyne planned to test the pumps separately, and a test program was started on May 1, 1975. Difficulties in testing the pumps made the program appear impractical. The testing failed to resolve any major problems and was abandoned by September 12, 1977. After that, the pumps were tested in the engine. When they failed, as a result of faulty pumps, seals, lines, or injector plates, as they did six times between March 24, 1977, and June 10, 1978, they damaged one engine after another. The space agency had to double the number of engines it originally ordered.

Pump problems became acute in 1976. The two main ones were a whirl mode—instability caused by vibration—in the rotor assembly suspension system and a lack of sufficient cooling for the bearings. These problems took time to identify because they were obscured by fires that broke out, especially in the oxygen pumping system; they also tended to mask each other.

The whirl mode was an old problem, virtually a tradition in reaction engine development. "Everybody who has ever made rotating machinery has fought whirl," said Lawrence O. Wear, deputy manager of SSME development at Marshall. By mid-1978, he believed the problem had been solved, although it was too early even then to be sure. "We just fought it until it went away," he said. "There are no cookbook recipes telling you how to get rid of it. Each whirl problem is unique."

The whirl problem was solved by stiffening the rotor assembly. That eliminated subsynchronous vibration, which had caused it.

Bearing failure had resulted in overheating, sometimes in fires. It was caused partly by an unequal load on some bearings and partly by inadequate cooling. It was solved by balancing the rotating parts to distribute the load on the bearings more evenly and by improving the cooling.

In spite of these problems, Marshall and Rocketdyne persisted in the belief that the engine design was sound. But experience told their engineers it would take trial and error to make it work.

Trial and error—the fail-and-fix process—was the implacable enemy of the concurrent-testing, success-oriented strategy NASA headquarters had adopted as a sop to Congress and the Office of Management and Budget. The OMB, the executive budget agency, functioned as financial overseer. The concurrent success-oriented strategy served its purpose of holding costs down. But it became evident in 1977 that the only strategy that would get this engine running was a failure-oriented one that would show the technicians what they were doing wrong. It was the high-cost way to go.

The Ad Hoc Committee

During 1978, an ad hoc committee of academic and industry engineers was formed to inspect the SSME development program at the request of the Subcommittee on Science, Technology, and Space of the Senate Committee on Commerce, Science, and Tranportation. The subcommittee expressed its growing concern about the persistence of engine test failures.

The Ad Hoc Committee that examined the "Technical Status of the space shuttle main engine" was a distinguished one, its members selected by the Assembly of

Engineering of the National Research Council. The chairman was Eugene Covert, Professor of Aeronautical Engineering at the Massachusetts Institute of Technology. In its first review, the Covert committee reported that the problems NASA and Rocketdyne were encountering with the SSME were "typical of those usually experienced in the early stages of a major technical advance."

In its early presentations to Congress, NASA had not specified that the SSME was a major technical advance. Instead, it had given the impression that although the engine would require some innovation, the technology for its development was well in hand.

The Covert committee questioned the lack of separate component testing. "No rocket engine using turbopumps has ever been produced without extensive development testing of the turbopumps first as components," the committee stated.

The report went on to suggest where NASA had failed. In order to avoid a careful, costly, and lengthy approach, in which each major component is developed and tested separately and in which alternatives are considered, NASA plunged ahead "with a fully concurrent, success-oriented strategy—a strategy based on the assumption that each piece of development hardware and its subsequent verification test will succeed on the first attempt."

"It may be," the committee added, "that this strategy has already saved time and money inasmuch as the quantity of hardware and spare parts have been reduced significantly. However, if or when malfunctions occur during the testing of the operational engine, new hardware may need to be designed, constructed and retrofitted, causing delays in the program. Indeed, such an eventuality has befallen the Shuttle main engine. . . . NASA and Rocketdyne have encountered some technical problems related mainly to the rotating machinery and the high pressure turbopumps which have led to engine test malfunctions and delays of test sequences."[2]

By the end of 1977, the failure-oriented strategy had asserted itself. Engine development was going along on the traditional test-and-fix pattern.

There were four expensive test failures that year. On March 24, a seal ruptured in the high-pressure oxygen pump. On September 8, the pump failed because of an overheated bearing. Friction caused a fire which damaged the engine. This problem was corrected by installing heavier bearings, equalizing the load on them, and redesigning the bearing carrier. On November 11, a turbine blade cracked in the high-pressure hydrogen pump in one engine, and in December the same failure occurred in another.

Repeated failure of a part was persistent. During 1978, there were seven major engine test failures. Three of them involved injection systems, in engines 0002 March 31, 0005 June 5, and 0006 November 3. The other failures were caused by nitrogen contamination in engine 0101 June 10, followed by an instrument failure in this engine July 18; a heat exchanger explosion in engine 0007 caused by a leak December 5; and the failure of a main oxygen valve in engine 2001 December 27 during a three-engine cluster test.

The valve failure was caused by oscillations in the liquid oxygen flow. The effect is called *pogo* because the liquid oscillates up and down. The failure resulted in a fire that damaged the engine and the test stand. The pogo effect was eliminated by redesigning the valve, but it caused a substantial delay in the engine testing program and also in the launch schedule.

Looking into the valve failure, the Ad Hoc Committee commented that it highlighted a shortage of spare parts and components; persistent shortages of parts and engines had been delaying the entire program. "This episode," it added, "underscores the earlier findings by the committee that parts and components need to be tested individually before they are assembled and re-tested as an engine system. If the main oxidizer valve had been mounted in a test stand and tested, the vibration and fretting might have been identified early in the test program."

It had become clear early in 1977 that as a result of engine problems there was no chance of launching the shuttle in March 1978, as the original schedule had set forth. The launch date was deferred a whole year, to March 1979. But in 1978, the continuing sequence of engine test failures made this launch date unlikely.

In the fall of 1978 John F. Yardley, who had suc-

ceeded Myers as chief of Shuttle development, advised the Senate subcommittee that "all program elements could be ready for a September 1979 first manned orbital flight . . . if all planned tests are successful."[3]

The Ad Hoc Committee regarded this forecast with doubt. In its view, the Shuttle was not in shape to be launched before March or April 1980.

During 1978, some progress on the engine had been made. A test of a three-engine cluster was run successfully for 100 seconds July 21. In September, a single engine was test fired three times at 100 percent of its rated power level for 520 seconds, the full duration of the launch to orbit. On October 30, an engine was fired for 823 seconds—the duration of one engine firing if the others failed, the launch was aborted, and the vehicle had to fly back to its base.

Close-up view of SSME system firing at the National Space Technology Laboratories test stand, Bay St. Louis, Mississippi.   (NSTL)

Remote view of main engine system firing test at the National
Space Technology Laboratories.   (NSTL)

Slowly, under the test-and-fix program, the SSME was
coming of age. By the fourth quarter of 1978, total engine running time had reached 34,810 seconds. NASA
required 80,000 seconds of test running time as a condition of launching the first orbital flight.

The President Appears

On October 1, 1978, the 20th anniversary of NASA
found the agency sliding from the pinnacle of success
with Apollo toward failure with the Shuttle. The idea that
the Shuttle could fail had never been considered seriously. Yet, six months after it was originally scheduled
to fly, its initial launch could not even be predicted with
certainty.

The anniversary was commemorated in Florida with a
visit to the Kennedy Space Center by President Jimmy
Carter, his wife, Rosalynn, and their daughter, Amy. It
was ceremonialized by the presentation of the Space
Medal of Honor to five living astronauts and to one post-
humously. The recipients were Neil A. Armstrong, first
man on the Moon and Professor of Engineering at the
University of Cincinnati; Frank Borman, commander of
the first circumlunar flight and president of Eastern Air-

lines; Charles Conrad, Jr., veteran of Gemini and Apollo,
who commanded the first crew to visit and repair Skylab; U.S. Senator John H. Glenn of Ohio, the first American in orbit; and Alan B. Shepard, now a Texas banker,
whose suborbital flight in Project Mercury opened the
U.S. manned space flight experience. The posthumous
award was made to Betty Grissom for her husband, Vir-
gil I. (Gus) Grissom, who perished in the *Apollo 1* fire in
1967.

This event was conducted in the cathedral-vast Vehicle Assembly Building, which had just been remodeled
for the assembly of the Shuttle. The VAB had been designed for assembling the Saturn-Apollo system. Much
of its steel had had to be cut away to make room for the
orbiter's 78-foot wingspread.

The occasion provided the president with his first opportunity to discuss the space policy of his administration. It would be a continuation of the low profile that
the United States had assumed in manned flight pending
development of the Shuttle, he said. U.S. manned flights
had ended with the Apollo-Soyuz joint flight in the summer of 1975. They would not be resumed until the Shuttle was launched.

"We will take our next great step into space with the
first flight of the Space Shuttle," Carter said, "which I
sincerely hope will be before my next birthday."

James Earl Carter, Jr., was born in Plains, Georgia, on
October 1, 1924. His visit to the space center marked
his 54th birthday. Would the Shuttle fly by his 55th?

"I have every assurance from those involved that there
will be no slippage in the present schedule as it is now,"
he said. It called for the first manned orbital flight of the
Shuttle on September 28, 1979.

Except for the Shuttle, there was no major development in Carter's policy. He had inherited the Shuttle from
the Nixon era. "I'm often asked about space factories,
solar power satellites, and such other large-scale engineering projects in space," he said. "In my judgment,
it's too early to commit the nation to such projects. But
we will continue the evolving development of our technology, taking intermediate steps that will keep open
possibilities for the future."

President Carter's visit to the Kennedy Space Center

had been intended to boost the space agency's fading image, but its effect was just the reverse. The "no new starts" policy he stated to space workers in the Vehicle Assembly Building that day was confirmed ten days later by a White House Fact Sheet on "U.S. Civil Space Policy." It stated in part, "It is neither feasible nor necessary at this time to commit the United States to a high challenge space engineering initiative comparable to Apollo."

What was so curious about the policy statement was the failure of the White House to recognize that the Space Shuttle Transportation System was, in fact, a high-challenge engineering initiative that would take the United States far beyond the range of Apollo. Despite his own engineering background, the president had seemed to overlook the potential of the Shuttle system for expanding the nation's economy in space. He seemed also unconcerned by the seriousness of the engineering problems that were delaying it.

Space workers summed up Carter's remarks in two words: No go.

Carter's no-new-starts space policy was sharply criticized by the chairman of the Senate Subcommittee on Science, Technology, and Space, Senator Adlai E. Stevenson (D-Ill.), and its ranking minority member, Senator Harrison E. Schmitt (R-New Mexico).

During hearings on the NASA appropriation for the 1980 fiscal year, Stevenson told NASA Administrator Robert A. Frosch, "I sense that the first priority of this administration is to study, study, and restudy. That's why this administration is such a small achiever."[3]

Schmitt, the geologist-astronaut who was propelled into the Senate from his *Apollo 17* mission to the Moon in 1972, called Carter's "no-go" space policy a form of "economic lunacy." He told the White House science adviser, Frank Press, "I can't tell from this [the budget] where the country's headed. There's nothing here but a negative commitment . . . to do nothing."

Schmitt charged the administration with lack of leadership under the "guise of fiscal restraint." It was forcing the nation to live off space investments made in the past, he said, and was "one of the most misguided efforts in the history of the country—perhaps in the history of mankind."[3]

November 9 . . . and So On

The twin test failures of engines 0007 and 2001 in December 1978 on the test stands in Mississippi doomed the September 28, 1979, launch date. Yardley set a new launch date—November 9, 1979. The Ad Hoc Committee remained skeptical. However, November 9 was confirmed by Frosch as the new "success-oriented" target date when he appeared before the Senate Committee on Commerce, Science, and Transportation on February 21, 1979.

Despite the engine setbacks in December, the SSME had achieved significant, long-duration runs, he said. The assessment of the vehicle as a whole was looking better. The solid rocket boosters had been fired successfully by the manufacturer in Utah four times, without any malfunctions. Now that the vibration tests were completed in Alabama, it was merely a question of perfecting the main engine and attaching the heat shield.

But perfecting the engine and applying the heat shield were tasks that were to remain incomplete for another 18 months, during which time the launch date was postponed again and again. Invariably, unforeseen problems kept cropping up to destroy the schedule.

The effect of delay and test failures on the funding of the Shuttle program came to the surface early in 1979, when Frosch went to Congress hat in hand to ask for a $185 million supplementary appropriation to the 1979 fiscal year budget. Development costs were escalating faster than the Shuttle directorate could estimate them, as new problems continued to plague the program.

Yardley explained to the Senate Subcommittee on Science, Technology, and Space that not only were engine test failures hiking costs, but unexpected difficulties had shown up in the thermal protection system, or heat shield.

The heat shield was a new invention. Instead of the old epoxy-resin shielding used on Apollo, a ceramic shield had been developed for the main body and wing of the orbiter. It consisted mainly of 31,000 silicate fiber tiles six and eight inches square, molded to fit the contour of the fuselage and wing surfaces. Other materials, such as pyrolized carbon, were used on the nose and wing leading edges, but the tiles covered most of the vehicle.

Yardley told the Senate subcommittee that the tiles had been difficult to manufacture "and the installation process has bugs in it." "As a result," he said, "we are behind on the installation of the tiles and Orbiter 102 [Columbia] will be delivered to the Kennedy Space Center without a number of tiles."[3]

The purpose of removing the orbiter from the factory in California before the heat shield was fully installed and completing the job in Florida was to speed up the work. NASA officials suspected it was lagging at the factory. In making the transfer, they were to encounter a Sisyphean problem that dwarfed even the engine failures in delaying the first orbital flight.*

Like the engine problems, the incipient tile difficulty was increasing the cost. Unless Congress appropriated the $185 million supplement, hopes for a launch in 1979 would go up in smoke, the subcommittee was told. As a former astronaut and manned flight advocate, Senator Schmitt sympathized with the request. He contended that the Shuttle had been seriously underfunded in 1972, by a factor of two, perhaps a factor of four. NASA's success-oriented strategy, he said, had served as a ruse to conceal the real technical problems and their cost. "When we first started space shuttle discussions, it was going to be an off-the-shelf vehicle," Schmitt recalled. "Well, it didn't turn out that way. Everything we're doing is new, leading-edge technical development."

The Ad Hoc Committee chairman, Eugene Covert, stated to the Senate subcommittee that the breakdowns on the test stand were "the price one should expect to pay for undertaking any hardware development program whose performance is beyond the art. The committee believes there is a paucity of development hardware in the program."[3]

One of the effects of the two major engine breakdowns in December 1978 was to illustrate the shortage of spare parts, the committee found.

Although failures should be considered a normal risk in any rocket development program, NASA's develop-

*Sisyphus was the mythological culprit whom the gods forced to roll a heavy stone perpetually up a steep hill, only to slip and let it roll down again every time.

ment strategy that brooked no failure and kept spares at a minimum escalated every test failure to a minidisaster. It reached the point that each time an engine failed on the test stand in Mississippi, the launch date slipped.

The Senate subcommittee chairman, Stevenson, blamed inadequate funding on the fiscal policy of the Office of Management and Budget. "I wonder what would have happened to Queen Isabella," he mused, "if she had an Office of Management and Budget? Those ships . . . never would have raised anchor."[3]

From this viewpoint, which Schmitt seemed to share, the problems of Shuttle development could be laid at the door of the federal bureaucracy. Ultimately, the responsibility rested on the Nixon and Carter administrations, which had attempted to build a reusable space transportation system on the cheap. With low priority, the Shuttle had become just another federal project, allowed to stumble along at its own pace.

Bricks Without Straw

By spring 1979, the Marshall Space Flight Center, which had charge of engine development, was ready to schedule the first full-duration (520 seconds) firing test of the Main Propulsion Test Article (MPTA) at the National Space Technology Laboratory (NSTL) in Mississippi. The MPTA consisted of the cluster of three engines housed in a mockup of the orbiter's rear fuselage.

NSTL was developed in the early 1960s as the Mississippi Test Center for Saturn rocket engines. It is located in a piny woods on the Pearl River near the Gulf of Mexico. The laboratory ranges over 13,500 acres and is remote enough from the nearest town, Bay St. Louis, to confine most of the rocket engine roar to the boondocks.

In Saturn-Apollo days, rocket stages were floated to the Center by barge from Michoud, an industrial suburb of New Orleans, where they were manufactured in a former aircraft factory. After testing, the big stages were shipped by water via the Gulf and around Florida to Port Canaveral on the east coast, serving the Kennedy Space Center.

NSTL is the largest rocket engine test center in the Western world. On the morning of May 15, 1979, the MPTA was prepared for firing in the 409-foot-tall test stand. It looked surprisingly large even though it was partially hidden by the structural steel work. I joined a small group of press observers who had filtered into the preserve from New Orleans, Washington, and New York, and we stood at the edge of a parking lot in the hot May sun waiting for word from the public address system.

At 10 A.M., word came. The test was postponed 24 hours. The day before, engine 0201, a development model, had failed in a test at a Rocketdyne–NASA test stand in Santa Susana, California. The test crew at NSTL wanted to know why.

What had that to do with the MPTA run? The fuel line had ruptured outside the nozzle on engine 0201. It was a section called the "steer horn" because it was shaped like the horn of a steer. In flight, such a failure would be disastrous. The test crew at NSTL decided to inspect the engines in the test stand for steer horn weakness.

After sundown, there was the usual assembly of engineers and technicians in a nearby village pub, where a tiny, agile waitress managed to serve a dozen crowded tables by dint of being in continuous motion, swooping, hovering, and fleeing like a sea gull.

I sat down at a table with three test engineers, whom I swore to identify only as Tom, Dick, and Harry. Now and then a fourth man leaned over from an adjacent table to make a comment. For this record, he shall be known as the Fourth Man.

We had a long discussion about the SSME, which I recorded on a tape recorder on the table. After the first few moments, no one paid any attention to it and the discussion waxed hot and intense. These people were pretty well fed up with shortages and delays. They had worked on rocket engines for years. Conditions now were as bad, as frustrating as they had ever seen. And it had been a bad day.

Tom spoke up: You want to know about shortages? All right. There are not enough engines. The reason for that is we've had as high a usage rate in this program as

any program I've ever seen. To put it bluntly, I guess we've torn up more hardware than I've seen on a development in a long time. But. . . .

Dick: Percentagewise, especially, for the amount of hardware we had available.

Tom: Yeah. First place, like he said, we've got less hardware to begin with. Maybe we've treated it a little rougher because we've gone at it harder and heavier. See, we've tried to put as much time on it sooner than we would in most development programs. As a result, we've overused hardware. The problem comes down to this: we're just hardware starved.

Harry: Supposedly, on the B-1 test stand, we have a system that simulates the orbiter, so far as the main propulsion system goes, fairly closely. But we have a little trouble in the fact we have one engine different from the other two. [It was in another stage of development.] We got to overcome that.

Tom (explaining): We started out with three MPTA configuration engines. We lost one back in December, in an explosion on the A-1 test stand. Okay? So they had to bring a flight-type engine into the program. It is a little more developed than the other two.

Dick: He means that we have two preproduction engines on the stand. Meaning that we had two engines not as far along in development as the flight type engine is supposed to be. The flight type engine has a different control system.

Author: All three engines now on the stand are not identical?

Harry: Right. We have an apple and two oranges. That's what we're going to have to run the rest of the MPTA program with. [Twelve MTA tests were scheduled in 1979.] Unless we have a hardware problem, all our static firings will be run with the same engines. If the NASA [launch] schedule holds, they'll be launching down there [at Kennedy Space Center] before we ever get through with our testing up here.

Fourth Man: I'll tell you what I think, and you write this down, Mr. Reporter. There's a lot of heat on this. It's a slow starter, a tough development program. My personal opinion is this government didn't put enough money in it. That's personal.

Dick: We got three major problems. One. Lack of money to begin with. Two. Trying to do component development at the same time we're trying to do full engine-up testing. Three. In all other programs where you had a new component, you got it to a certain level of maturity before you put it into the system. We didn't do that. I'm saying I don't know why.

Tom: Two reasons, I think. One is lack of money. The other is component. . . .

Harry: My personal opinion is . . . they didn't start out like all normal programs do. They had a time constraint, a money constraint, or something. They figured they could bring it off anyway. But I think they underestimated the difficulties. Or they overestimated the state of the art. Every engine Rocketdyne's developed has had a [rotating machinery] problem. You've 400 pounds of machinery coming up to speed at 40,000 revolutions per minute in four seconds. That's massive. Pressures and temperatures are higher than ever.

Tom: In Apollo days, you had several contractors spread out over pieces of hardware. NASA monitored each individual part of the program. In this program. . . . we've got a lot more supervisors than we had before. In my opinion, they get in my knickers. They slow us down. Fourteen people telling you to do something. We don't even go to the bathroom unless we have a committee of ten. It's a conservative philosophy.

Fourth Man: We ain't bitten off more than we can chew. The engine is good. Basically good. The damn thing will work. It needs more money and patience. We don't have any problems on that engine that aren't solvable.

Dick: I don't know that I'd want to crawl into that thing [the orbiter cockpit] the way it is. More than anything else, it needs time. That's what nobody wants to give it. Nor money either.

We broke up then. I moved to pay for the beer, but the little waitress flying by said it had been taken care of.

Who? I asked as she soared by again.

Why, that gentleman with the necktie, she said.

I didn't remember seeing anyone wearing a necktie. It was't that type of place.

The next day, the test was postponed another 24 hours, and later, still another 24. It slid into June. On June 1, inspectors found a crack in the nozzle of one of the pre-production engines on the stand. The test was postponed again.

On June 12, the MPTA was ignited and the engines ran at 100 percent of full rated power for 55 seconds. They cut off at about one-tenth of the time they were supposed to run because of excess vibration on engine three.

On July 2, a second attempt to run the full-duration test was stopped at 18.5 seconds. Hydrogen was leaking from the fuel valve on one engine. Fire broke out from an accumulation of hydrogen, but there was no explosion.

When the fire was put out with nitrogen and water, it was found that the aft heat shield on the orbiter mock-up was damaged and the insulation of the external tank was charred.

The ill-luck MPTA test was postponed until October. That meant that the first orbital flight of *Columbia* would slip into 1980.

6

The Great *Skylab* Rescue Mission

The main engine failures were the harbingers of hard times ahead. They were to continue right up to the launch of the first manned orbital test flight and beyond.*

As failures persisted and launch dates slipped, Senate and House space committees came to realize that the Space Shuttle Transportation System was a major technological challenge; that although the vehicle was designed to fly only in low Earth orbit, it was considerably more sophisticated than the Saturn-Apollo system—if less dramatic.

By mid-1978, it had become clear to the spacework community that the Shuttle challenged the capacity of American technology. It was not a challenge that an industrial nation could afford to lose, especially one that

had committed $21 billion a decade earlier to land the first men on the Moon.

Had the extent of the challenge been realized in 1970, I doubt that the project would have been undertaken at all. That time of social troubles and reappraisal of national priorities did not encourage new programs.

The realization that the Shuttle was no small step up from Apollo came as a shock to the entire NASA bureaucracy. This was dramatized in an unexpected way by a sudden crisis precipitated by the end product of Apollo-Saturn—era technology, the big space station, *Skylab*.

Skylab was falling out of orbit, where it had been expected to remain until the mid-1980s. Its premature descent had first been disclosed by the British and later confirmed by the North American Air Defense Command (NORAD).

The largest man-made object in orbit, *Skylab* had been

*There was no letup. Hydrogen leaks in the main engines of the *Challenger*, the second flight orbiter, delayed its launch three months in 1982–1983.

Skylab, NASA's first space station, was the largest manned orbital vehicle built in the 1970s. This photo was taken with a 70mm hand-held Hasselblad camera from the Apollo spacecraft used to ferry the second crew to and from the station; the crew flew around the station prior to docking with it July 28, 1973. The windmill structure to the right, with solar panels outstretched, is the Apollo Telescope Mount, a solar observatory. Below it is the multiple docking adapter, with a circular docking port on the forward end. The main part of the structure is the Orbital Workshop, from which extends a single wing of solar cells.   (NASA)

left as a derelict early in 1974 when the last of three crews departed after inserting it into an orbit that was supposedly high enough to keep it aloft for another decade.

It was a huge derelict, 86 feet long, weighing about 170,000 pounds. Late in 1977, it became apparent to tracking services in the United States and Britain that *Skylab*'s orbit was decaying much sooner than expected.

This seemed to be due to an unanticipated increase in activity on the sun, manifested by an increase in sunspots. It bombarded the top of the atmosphere with high-energy nuclear and subnuclear particles. The heating effect of the bombardment caused the atmosphere to rise; tenuous wisps of it reached *Skylab*'s orbit and began to slow down the vehicle, causing it to lose altitude. As it did so, as it sank gradually into denser atmosphere, its rate of descent accelerated.

Satellite tracking data gathered by NORAD at Colorado Springs predicted that *Skylab* would reenter and crash in or before 1980—four years early.

The prospect of a premature *Skylab* reentry had not appeared to be a matter of deep concern to NASA until the widely publicized crash of the Russian spy satellite *Cosmos 954*. It came down in western Canada January 24, 1978, leaving a trail of radioactive debris from its power generator. Its reentry was also premature and unexpected.

Skylab had no radioactive material aboard, but it had tons of metal hardware which might survive the heat of reentry as the space station broke up. The fall of big, heavy pieces, as well as of thousands of smaller ones, posed a threat that seems to have been overlooked when *Skylab* was abandoned in 1974.

Launched May 14, 1973, from the Kennedy Space Center, Florida, at an inclination of 50° from the equator, the space station passed over most of the planet's centers of population—between 50° north and 50° south latitude. Although 70 percent of the track was over water, there was no guarantee that that pieces of the station would miss populated areas in an uncontrolled plunge.

NASA's *Skylab* directorate had expected the station to remain aloft until 1984, when the Shuttle would be flying. Then, perhaps, the Shuttle could be used to boost it to a storage orbit or control its plunge into the ocean. Now NASA was confronted with a dilemma. What could it do about an early reentry? Could the Shuttle be made ready in time to effect a rescue?

Such a mission could be a dramatic affirmation of the Shuttle's utility. It would "rescue" not merely *Skylab* but also NASA, by averting the embarrassment, lawsuits, and international complications which the fall of *Skylab* debris in a populated area could generate.

In March 1978, crews for the first two orbital test flights of the Shuttle were announced. John W. Young, 48, was named commander and Robert L. Crippen, 41, pilot of STS-1.* Fred W. Haise, 44, and Jack R. Lousma, 42, were named commander and pilot, respectively, of STS-2. With this announcement, NASA disclosed it was considering using the second flight to hitch a propulsion unit to *Skylab* and avoid an uncontrolled reentry of the big space station.

It was a bold idea, in keeping with the boldness that had made Apollo so spectacular and successful. During the spring 1978 Space Congress, an annual technical review of space activity sponsored by the Canaveral Council of Technical Societies, Fred Haise said that he and Lousma were in training for the rescue mission.

Although the specifics of the mission were not spelled out at the time, the plan was to make a rendezvous with *Skylab* and maneuver a radio-controlled booster rocket to dock with it. The orbiter would then move away, and the rocket would be fired by a radio signal.

The decision to boost *Skylab* to a higher orbit or use the rocket to deorbit the station and send it diving into the Pacific Ocean would be made by NASA at the time of the flight. The prospect of a *Skylab* rescue on the second orbital test flight added a certain glamor to the otherwise glum record of Shuttle development.

Two preconditions had to exist. One was that the second orbital flight had to be launched by fall 1979; the other was that the radio-controlled booster rocket had to be made ready in time.

NASA's Cabin in the Sky

A word about *Skylab*. The station was developed from Gemini, Apollo, and Saturn rocket parts. It was an ad hoc effort designed to generate more advanced versions of a space station as a terminal for the Shuttle.

NASA announced the project July 22, 1969, as *Apollo 11* was returning to Earth from the first landing on the Moon. *Skylab* was not a single structure but an agglomeration of several, new and old. The main body, the Or-

*Space (Shuttle) Transportation System.

bital Workshop (OWS), was a cylinder 48 feet long and 22 feet across, with a volume of 9,550 cubic feet. It was the remodeled shell of the propellant tank of the Saturn 5 third stage.

The OWS contained sleeping quarters, a kitchen, toilet, and shower for a crew of three, plus work space for experiments. NASA publicity compared it in spaciousness to a five-room house.

Attached to the Workshop were an airlock, for entry and exit; a docking port, where Apollo could be docked to discharge or take on the crew; and a solar observatory with one of the most powerful arrays of instruments ever sent into space, to observe the Sun. Unfortunately, the data obtained from the solar observatory failed to predict the increase in solar activity that caused the fall of *Skylab* years earlier than expected.

The entire *Skylab* assembly, with Apollo docked, was 118.5 feet long and weighed 100 tons. Total working space, including pressurized volume in the airlock, the docking port, and Apollo, was 12,398 cubic feet, or 347 cubic meters. It was more than three times the habitable volume of the Soviet space station, Salyut. Compared to Salyut and the Manned Orbital Laboratory proposed by the U.S. Air Force, *Skylab* was a palace.

Skylab was occupied a total of 171 days by three crews, which were shuttled up to it and back to Earth in Apollo spacecraft, boosted by Saturn IB launchers. The first crew, consisting of Charles (Pete) Conrad, Jr., Paul J. Weitz, and Joseph P. Kerwin, M.D., spent 28 days aboard the station (May 25 to June 22, 1973). Most of their time was spent rigging a sunshield and making repairs to the OWS, which had been badly damaged during its launch. The second crew, Alan L. Bean, Owen K. Garriott, and Jack Lousma, continued repairs and spent 59 days in the station, from July 28 to September 25, 1973. The third crew, Gerald P. Carr, Edward G. Gibson, and William R. Pogue, set a U.S. flight record of 84 days from November 16, 1973, to February 8, 1974. Before leaving the station, they jockeyed it into an oval orbit of 282.3 by 270.25 nautical miles (324.6 by 310.6 statute miles), using the Apollo service module's space engines. Expecting the station to remain in a safe orbit for nine or ten years, by which time the Shuttle should be flying to move it if nec-

essary, Mission Control at Houston sent a radio signal to shut off *Skylab*'s power, and the station was abandoned.

There was strong sentiment at the Marshall Space Flight Center for rehabilitating *Skylab* and using it during the Shuttle era. That view was not shared at headquarters. William E. Schneider, who had directed the program, pointed out that the station had been designed to operate for only nine months. If it were to be used for a longer term, it would have to be rebuilt in orbit—a difficult, costly undertaking.

So *Skylab* drifted. As the years passed, it seemed to recede into oblivion until NASA was alerted that the station was coming down early.

Retrieval

The premature, uncontrolled reentry of *Skylab*, following the Cosmos crash in Canada, forced NASA to make a contingency plan to control the descent of the station. The space agency's initial move was to try to regain control of the station, after having allowed it to drift for four years as a derelict.

Control might be reestablished if the station would respond to radio signals that activated the electrical system. If power could be restored, the control gyroscopes and attitude control thrusters could be operated. The station could be turned into a "low drag" attitude to minimize air resistance. Such a maneuver would extend its stay in orbit.

Next, the agency called for development of a remotely controlled rocket stage that could be carried into orbit by the Shuttle and hitched to *Skylab*'s docking port. It would give NASA the option of boosting the station to a "safe" orbit or of guiding it to an ocean splashdown.

The rocket stage was called the teleoperator retrieval system, or TRS. It was to be built as a general-purpose, remotely controlled, reusable upper stage that could be used to maneuver payloads to higher or lower orbits, to retrieve satellites for servicing, inspection, or repair, or to deploy experiments. Its first use would be to retrieve *Skylab*—if the Shuttle could be launched in time.

The Denver Division of Martin Marietta Corporation

For a while, NASA hoped to control the descent of *Skylab* by docking with it a propulsion system operated from the Shuttle. This is an artist's drawing of the unit, a Teleoperator Retrieval System, that Martin Marietta Aerospace was to develop for the Marshall Space Flight Center. The docking mechanism is shown on the forward (left) end of the device. The project was cancelled when it became apparent that *Skylab* would come down to Earth before the Shuttle could be launched.

(NASA)

was awarded a contract by NASA to manufacture a TRS for $35 million. The vehicle was to be ready for flight by May 1, 1979. A docking device was to be mounted on its forward end, along with two television cameras that would help the Shuttle crewman who was operating the rocket to steer it into *Skylab*'s docking port. The docking device was the same as that used by Apollo in the 1975 joint mission with the Soviet Union's Soyuz spacecraft.

When lifted out of the orbiter's cargo bay by the remote manipulator arm, the TRS would be set on a course for *Skylab* under the control of the astronaut operating it from a special console. When it was docked with *Skylab,* the crew would await instruction from Washington for the next event.

If the decision was to keep *Skylab* aloft for possible future use, the crew would be ordered to boost it to a higher orbit with two 13.5-minute burns of the TRS rocket engines. If the decision was to give *Skylab* the deep six, it would be deorbited with one long burn of 27 minutes and reenter at a point where it would fall into the Pacific.

The astronaut most likely to control the TRS from the orbiter was Lousma, the pilot on the second orbital flight of the Shuttle. Having been a member of the second crew to visit the station in 1973, he presumably knew it well.

In early February 1978, Houston was advised that *Skylab* was likely to begin its reentry by late summer 1979—although there was a possibility it would remain aloft until sometime in the first quarter of 1980. This advice was based on data from NORAD, the Smithsonian Astrophysical Observatory, and the Swiss Federal Observatory. *Skylab* was then orbiting the Earth at an altitude at 220 nautical miles. It had dropped 50 miles below its 1974 perigee of 270.25 nautical in 1974, or about 10.25 miles a year. At first, the descent had been gradual; now it was accelerating.

In March 1978 a team of engineers from the Marshall and Johnson centers was assembled to attempt the first step in gaining control over *Skylab*—reactivating its attitude control system.

The team managed to reestablish radio contact with the station from NASA's Bermuda tracking station, and to continue the contact from the space agency station in Madrid. To NASA's surprise and relief, the electrical system was working as long as the solar panels were in direct sunlight. To keep them there, the team maneuvered the station into a solar inertial attitude so that the solar cells were fully exposed to sunlight on the day side of the orbit.

By early June, the batteries were sufficiently charged so that *Skylab* could be placed in a position where its docking port was forward and its long axis was parallel to the ground. Now it was flying like a fat torpedo. This attitude was referred to by the team as the "end on velocity vector."

On July 9, alas, the power system failed. The charging process was causing the batteries to overheat to the point where a safety switch turned them off. By July 19, controllers found that by charging sets of batteries in rotation, they could avoid overheating them. Power was restored.

At Marshall, sentiment for resuscitating *Skylab* surged into the open. A press release was issued, suggesting that "the large living quarters and crew accommodations aboard *Skylab* would be a welcome adjunct to Space Shuttle and Spacelab missions, involving extensive mission equipment and long mission durations. Useful additional experiments might be conducted with *Skylab* instruments; also the possibility for new experiments, missions or demonstrations would be realized with Orbiter and Spacelab docked with *Skylab*." The Marshall folk went on to speculate that such an array of hardware in orbit could provide the platform for assembling large space structures, such as solar power satellites and other "public service" facilities.

As engineers at the Marshall and Johnson Centers struggled to prolong *Skylab*'s flight, Haise and Lousma continued training for the STS-2 flight, as did their backup crew, Vance D. Brand and Charles G. Fullerton. Except for Fullerton, who had not flown in orbit, it was a spacewise group. Haise had been lunar module pilot on *Apollo 13*'s abortive mission to the Moon; Lousma had spent 59 days on *Skylab;* Brand had been a member of the Apollo crew on the 1975 joint flight with the Russians.

Discussing the rescue operation briefly, Haise said that docking the TRS with *Skylab* would be conducted with conventional rendezvous and docking procedures. He anticipated four days of phasing maneuvers with the orbiter to accomplish the rendezvous with *Skylab*. Then the TRS would be deployed and sent cruising to its target.

During the summer of 1978, a tracking station at Santiago, Chile, was added to the *Skylab* control network. The network by then included Goldstone, California, in addition to Santiago, Bermuda, and Madrid.

During October 1978, controllers at Houston turned *Skylab* around, front to back. This was done in order to expose to the Sun a gyroscope that was beginning to freeze.

The low point, or perigee, in *Skylab*'s orbit continued to drop. As a result, every time the big station passed through perigee, it met increased air resistance. Even though the atmosphere at *Skylab*'s altitude was so sparse

as to constitute a near-perfect vacuum, there were enough molecules of air to slow the vehicle, little by little.

Orbital mechanics tells us that a change in velocity at perigee, no matter how slight, produces a change in the altitude of apogee, the high point in the orbit. Consequently, as the vehicle was slowed by increasing friction at perigee, its apogee dropped. Each time around, it came gradually lower and lower, like Edgar Allan Poe's pendulum, the works of civilization prostrate below.

Day after day, this process continued. Day after day, observatories and radar stations measured the change. Despite all of the efforts of engineers, *Skylab* was falling, the last artifact of the Apollo era. There was no telling where it would reenter. Although NASA predicted it would break up in the lower atmosphere, no one could predict where large, heavy pieces of the first space station would land.

Day by day, week by week, the inexorable descent of *Skylab* made the delay in bringing the Shuttle to the launch pad more painful. Lack of initially adequate funding resulted in the stretchout of the development program. A product of overconfidence, underfunding now contributed to the failure to rescue *Skylab*. Its uncontrolled plunge for a time threatened an unparalleled disaster for the habitats of mankind between 50° north and 50° south latitude—*Skylab*'s ground track. Had Shuttle development adhered to its original schedule, a *Skylab* rescue mission would have been eminently feasible on STS-2, the second flight of *Columbia*.

But even as John Yardley had marked the September 28 launch date on the calendar, he had qualified it as only "probable" when he appeared before the House Subcommittee on Space Science and Applications. It was more likely, he said, that the first flight would not take place before the fourth quarter of 1979. Yet a September 28 launch lingered as a faint hope until the dual test failures of engines 0007 and 2001 in December 1978 finally erased it.

Headquarters set a new tentative launch date of November 9, 1979. Like the earlier dates, it depended totally on the absence of surprises.

But they kept on showing up. NASA and Rocketdyne

engineers were still dealing with an immature rocket engine system during this period. There were no shortcuts to bringing it to maturity.

The Ad Hoc Committee of university and industry engineers that had been appointed by the National Research Council Assembly of Engineering to review Shuttle engine development at the request of the Senate remained skeptical about any launch date in 1979. The Committee held to its prediction that no initial launch was likely before the first or second quarter of 1980. NORAD, meantime, predicted that *Skylab* would reenter the atmosphere and crash before then—possibly as early as July or August 1979.

When it became evident that Shuttle development delays had put a *Skylab* rescue mission out of the question, NASA abandoned it. The agency announced December 19, 1978, that it was discontinuing preparation for the mission because of its "limited potential for success." The earliest a TRS mission could be flown would be April 1980, a spokesman said.*

Production of the TRS was halted. About $20 million of unexpended funds were transferred to Shuttle manufacturing, where they were sorely needed.

In January 1979, controllers maneuvered *Skylab* into a "solar inertial" attitude, so that its solar panels faced the Sun. The station then flew perpendicular to the ground. Drag was high. While this attitude would reduce orbital lifetime, it would keep the batteries charged so that Houston could maintain control.

A new *Skylab* game was devised. By changing the flight attitude from time to time, controllers believed they could control the point of reentry and hence that of landing. They could change the impact area of *Skylab* debris if a population area were threatened. By maneuvering the station so that it flew front end forward and parallel with the ground, controllers could extend the flight by hours, or even days. By turning the vehicle so that it flew sideways, in a high-drag attitude, they could reduce the flight time.

*Haise resigned from NASA in 1979. Engle and Truly were assigned to STS-2 and Lousma and Fullerton to STS-3.

If the space agency could win the game, it could avert a calamitous impact of heavy metal in a populous area. The technique of controlling a falling satellite by attitude control was new, untried, uncertain. It suggested desperation. The agency had to do something, and this was all it could do, now that it had lost the Shuttle rescue option.

The name of the new game was "drag modulation." At a meeting at Houston in late spring, the NASA directorate appointed Richard G. Smith, a former Marshall executive, as coordinator of this effort and chief spokesman for the agency during the last, crucial days of *Skylab*'s plunge. Smith knew *Skylab* as well as anyone. A big, ruddy, genial North Carolinian, he was the ideal choice for this sensitive job. The sensitive part of it was dealing with the press.

By late spring, a drag modulation technique had been worked out that would increase or decrease *Skylab*'s orbital lifetime by one or more revolutions around the Earth, as long as Houston could maintain attitude control. Thus, if the station appeared headed for reentry at a point that would allow pieces of it to spray over a metropolitan area, its attitude could be altered to cause it to reenter at another point, where the remains would splash down in the Atlantic, Indian, or Pacific Ocean.

Four attitudes were available: solar inertial, which kept the batteries charged but produced the highest drag; the end-on position, which produced the lowest drag; tumbling, which produced an intermediate drag force; and a sideways flying position called "torque equilibrium." Houston prepared to play this celestial billiard game in June.

The Cosmic Shotgun

Starting on June 19, 1979, controllers maneuvered *Skylab* into the fourth attitude, which was deemed necessary to prevent it from tumbling prematurely as it encountered increasing air resistance. In this torque equilibrium attitude, the station was flying sideways, with its long axis parallel with the horizon—a high-drag but stable position.

A stream of reports from NORAD indicated the end was near. *Skylab* was at 166 nautical miles, and reentry was expected between July 7 and 25.

By the end of June, a swarm of engineers and mathematicians were laboring over calculations to determine how best to control reentry and impact. It was the first exercise of this kind ever attempted. Could the plunge of this 85-ton derelict be managed from the ground? No one could be sure, but more than 300 men and women at NASA centers and at tracking stations around the world were working around the clock to try it.

A study by the Battelle Memorial Institute showed that as much as 25 tons of metal would reach the ground in 500 pieces or more. The largest probably would be the 4,950-pound airlock shroud and the 39,600-pound lead-lined film vault, which Battelle figured might fall as one piece with an impact velocity of 393.6 feet a second. In addition, there were six 2,650-pound oxygen tanks on board and a 14,960-pound bulkhead that might survive reentry heating.

Briefly, NASA engineers studied the effect of using missile warheads to break up the station in orbit and thus control the time of the spread of debris. But they concluded this tactic would only increase the spread of debris and rule out any possibility of controlling reentry.

The 500 pieces or so of metal that Battelle predicted would come through reentry would be spread over a "footprint" 4,000 miles long and 100 miles wide. *Skylab* had become a shotgun, aimed at civilization. However, NASA publicists issued reassuring handouts advising the public that the chance of personal injury or property damage from debris was no greater than that from meteorites.

Nevertheless, public reaction was escalating as the media built up a doomsday story. The reaction was mixed. There were a few letters to the editor in newspapers asking why NASA had allowed this situation to develop. A fringe registered a "silly season" response, exhibiting *Skylab* T-shirts and caps with bull's-eye targets. The impending fall of *Skylab* set off no general panic; it seemed to be too incredible to take seriously.

The deepest worry was seemingly confined to NASA and the State Department. If injury or damage occurred,

NASA would be raked over the coals and the government would be beset with damage suits. Allowing pieces of hot metal to bombard other countries was a poor way of impressing them with America's technological prowess. All the elements of a national and international debacle were in the making as *Skylab* kept falling. NASA had never looked so bad.

On July 4, the space agency prepared for the reentry on July 10 or 11. Dick Smith began his vigil as Skylab Task Group coordinator at NASA headquarters. His task: to find out what was going on from NORAD, Marshall, and Houston; take part in the course of action; explain it all to the news media; and reassure everybody.

On July 10, as the moment of truth approached, Smith held a news conference. He estimated that the solar panels would come off at 65 nautical miles altitude, the solar observatory would break loose shortly after that, and the station would come in as two big pieces, each with its own point of entry. Below 50 miles, the pieces would get hot; below 45 miles, they would break up into hundreds of little bodies, each with a different size, shape, mass, and entry trajectory. All of this was speculative. No one really knew what would happen.

Early on the morning of July 11, Smith announced that *Skylab* would be nudged out of the torque equilibrium attitude when it reached 80 nautical miles altitude and allowed to tumble. That maneuver would lengthen its time aloft somewhat.

NORAD predicted a reentry on revolution number 34,981 over southern Canada, just north of the United States border. The estimated impact area was heavily populated. By tumbling *Skylab*, controllers hoped to extend the orbit so that it would fall into the South Atlantic or Indian Ocean.

At 2:49 A.M. eastern daylight time on July 11, *Skylab* reached 80.8 nautical miles. It was estimated to be nine hours from reentry. Commands were radioed from Madrid to start it tumbling. At 4:19 A.M., the Madrid tracking station confirmed that *Skylab* was tumbling over one axis. The maneuver had used 2,000 pounds of thruster gas.

At 5:23 A.M., *Skylab* was reported to be at 78 miles. NORAD predicted reentry between 10:41 A.M. and 1:01

P.M. Midpoint of the descent pattern would be in the Atlantic, several hundred miles east of New York City.

At 10:02 A.M., NORAD sharpened its reentry prediction to the period between 11:09 A.M. and 1:09 P.M. It would reenter over the Atlantic and splash below the equator. It continued to tumble ever lower.

NORAD issued its final forecast at 11:25 A.M. Reentry was refined to the period between 12:01 and 12:53 P.M. eastern daylight time. Smith and his colleagues at headquarters were able to breathe a little easier, or so they thought. It was certain now that *Skylab* would clear North America. Once it had done so, the coast would be clear all the way to Australia. *Skylab* was headed for the deep six.

At 11:47 A.M., *Skylab* passed over North America and was acquired by the Bermuda tracking station. The experts had expected it to break up at that point, but the radar showed it was still in one piece.

The station continued descending along the South Atlantic track and at 12:01 P.M. it was picked up by the isolated tracking station on Ascension Island. The solar panels were still intact, and the vehicle was higher than expected. It should have broken up at Ascension, according to the calculations.

Happy grins in Washington and Houston faded. However, while still within range of Ascension, telemetry signals from the airlock module ceased. It appeared that *Skylab* was beginning to break up at last.

The ground track passed near Kerguelen Island on the rim of the Antarctic Ocean and then swung northeast toward Australia. The big interplanetary tracking and data collection station at Muchea, Australia, was not operating. The next telemetry-radar acquisition site was an Air Force station on Kwajalein Island at 12:45 P.M.—if the vehicle was still in the air by then.

But it never reached Kwajalein. Just past noon, NASA began receiving long-distance telephone calls from Perth reporting spectacular fireworks in the sky. Perth is on the southwest coast of Australia, and its inhabitants had turned on their lights for John Glenn to hail the first American orbital flight as it passed over their city February 20, 1962.

It was 1 A.M. in Perth now, and the folk there were seeing another celestial visitor—*Skylab*. Its pieces, bright and hot, were raining down around Perth and the southwest coast. NASA's drag modulation effort had not worked as well as hoped.

Excited callers reported seeing 50 to 100 bright flares in the sky, all colors, like fireworks. It was a magnificent display. Residents of the towns of Albany, Esperance, and Kalgoorlie were deluging switchboards at radio stations, police stations, and fire stations with reports and questions. Enterprising radio news teams were telephoning NASA in Washington to relay the reports. *Skylab* was coming down on Australia like confetti, from the coast to inland over the sparsely populated desert outback.

An airline pilot who landed at Perth called NASA to say he had seen very large pieces breaking up into fiery shards while he was still aloft at 29,000 feet.

When all of these reports were digested, Charles S. Harlan, the *Skylab* flight controller at Houston, issued the understatement of the space age: "We assume that *Skylab* is on the planet Earth, somewhere."

NASA and the State Department anxiously waited throughout the day for complaints of damage, or worse. None came. By late afternoon, the space agency directors were breathing easier.

Had *Skylab* continued for one more revolution, its ground track would have passed through Minneapolis–St. Paul; Columbus, Ohio; and eastern Virginia before moving out to sea at Cape Hatteras, North Carolina.

Although several residents of Perth complained about NASA dumping its refuse on Australia, the NASA inspection teams that spent weeks there examining the impact area were jovially received.

Australians in the Perth area will long remember the night that *Skylab* fell. Certainly 17-year-old Stanley Thornton, Jr., who collected 24 charred pieces of *Skylab* from his backyard in Esperance, won't forget it. Stanley, his parents, and his girlfriend were flown to America by a Philadelphia furniture man, and the young man received a $10,000 reward for finding the first authentic piece of *Skylab*. The reward had been posted by the San Francisco *Examiner*. In addition, it was reported, he received $1,000 from the furniture man.

Of course, the visitors toured Disney World and the Kennedy Space Center. Stanley told a television inter-

Members of the *Skylab* reconnaissance team examine pieces of the space station they brought back from Australia for analysis at the Marshall Space Flight Center. Left to right: Billy Adair, electronics and trajectory specialist; Joseph M. Jones, public affairs chief; Dr. Ray Gause, materials specialist; and William (Ozzie) Harrison, structures specialist. The samples were analyzed for effects of more than six years' exposure in space.
(NASA)

viewer he wasn't terribly impressed, however. And when would they get the Shuttle up?

So ended the saga of *Skylab,* the grand hotel of space stations. Its premature reentry highlighted the delay in the development of the Shuttle as nothing else could have done. But it taught NASA several lessons. One was never to put a large object in orbit again without some means of controlling its altitude. Another was the new technique of drag modulation as a means of controlling reentry.

Analysis of the debris recovered in southwest Australia showed that the station had not disintegrated until it had reached an altitude of 10 miles. That was six times lower than calculated. Debris was scattered over an area 40 miles wide and 2,400 miles long—a smaller footprint than calculated.

During the autumn of 1979, a 180-pound piece of aluminum that might have been the door to the film vault turned up. Two large oxygen tanks and a pair of titanium spheres containing nitrogen were found new Rawlinna, about 275 miles east of Perth. The NASA inspectors concluded that the debris spread ended about 500 miles northeast of Rawlinna.

Smith, who had coordinated the last act of the *Skylab* drama with skill and efficiency, was subsequently appointed director of the Kennedy Space Center. On the afternoon of July 11, after it was all over in Australia, he went home to take a nap before dinner. He had been working under heavy pressure more than 24 hours.

It was marvelous that anyone could just doze off after the strain of the last days of *Skylab.* "Well, I did," Smith recalled a few months later. "And I woke up with the most god-awful headache anyone ever had."

With *Skylab* scattered in the Indian Ocean and over the Australian outback, the Apollo era came to an end. Its last hurrah was a sigh of relief.

7

Shield of Sand

By spring 1979, it was unlikely, insiders conceded, that the first manned orbital flight of *Columbia* could be launched before 1980. NASA's engine test requirements could not be met before then without exceptional luck. Also, installation of the heat shield on *Columbia* was lagging at the Rockwell plant at Palmdale, California.

Nevertheless, NASA headquarters held to its target launch date of November 9, 1979. Ostensibly to meet it, headquarters ordered *Columbia* transferred to Florida. Agency managers believed that heat shield installation could be speeded up at the Kennedy Space Center, where *Columbia* could also be made ready for flight at the same time. The work would be done in the Orbiter Processing Facility (OPF), a big hangar designed to refurbish orbiters between missions.

As preparations went forward in California to ferry *Columbia* across the country atop NASA's 747 carrier airplane, a new problem appeared. The ferry flight was scheduled to take off from Dryden March 9, but an over-

sight delayed it a day. The mate-demate device that had been used to hoist the *Enterprise* onto the jumbo jet had to be modified to fit *Columbia*. This adjustment held up the transfer until March 10. It was the start of a comedy of errors that was to put off arrival at Kennedy until March 25.

Scene one was the trial hop Dryden ordered to check the aerodynamic compatibility of *Columbia* with the 747. The transfer would be a long haul, and the orbiter was in an unfinished state. It lacked rocket engines in the tail and one-fourth of its heat shield was missing. A tail cone was installed to smooth air flow, but the vacancies in the heat shield had to be filled temporarily to avoid turbulence during the flight.

As mentioned earlier, the main part of the heat shield consisted of 31,000 ceramic tiles. They varied in thickness depending on the intensity of the frictional heat they were expected to encounter during reentry into the atmosphere.

About 23,200 tiles had been glued onto the aluminum skin of the orbiter at the factory, leaving 7,800 to be installed at Kennedy Space Center. In order to fill in the gaps where tiles were missing, "dummy" tiles had been inserted temporarily, held in place by adhesive tape. At Kennedy, they would be pulled off and replaced by permanent tiles.

After noon on March 10, NASA's Boeing 747, with *Columbia* perched on top, rolled down the Dryden main runway for the test hop. A groan came from the tower. The dummy tiles were cascading off the orbiter like confetti. As the jet lifted, tapes flapped in the wind and ripped off more than 100 of the permanent tiles. Some of them just seemed to fall off. The permanent tiles were not sticking. Either the bonding method had failed or the permanent tiles were splitting.

If the tiles came off during launch, the flight would have to be aborted and the orbiter would return to base. If they came off in orbit, the crew faced incineration when the orbiter reentered the atmosphere, at temperatures up to 2,700° F.

Fitzhugh Fulton, the 747 pilot, called Dryden Control to please get some people with brooms out on the runway to clear away the broken tiles so that he could land.

There was no joy in Mudville, for the thermal protection team had struck out. When *Columbia* was rolled back into the hangar for inspection, it was found that 4,800 dummy tiles and 100 permanent tiles had come off.

A force of 100 men and women was mobilized at Dryden to reinstall the dummy tiles. This time, they were glued on through holes drilled in the corners. The bonding agent required 60 hours to dry. On March 18, the new method of putting on the temporary tiles was tested on a T-38 jet aircraft. It was pronounced okay and *Columbia* was declared ready. Next day, the curtain rose on scene 2.

Remated with the 747, the orbiter was scheduled for the flight on March 19, with arrival at Kennedy the next day. But that was not to be. A spring storm roared in from the Pacific northwest, delaying the takeoff until March 20.

The 747 and its payload did get off then, but caught up with the storm in Texas. Unable to land at San Antonio for scheduled refueling, the 747 crew put down at Biggs Army Air Field at El Paso to sweat out the weather. The crew stayed on the ground during March 21 and managed to reach San Antonio in wild weather on March 22. The storm was traveling along their flight route, and they had to stay behind it.

They flew to Eglin Air Force Base on the Gulf Coast of Florida on March 23, remained overnight, and on the morning of March 24 started the engines for last lap to Kennedy—a short hop across the Florida panhandle and the peninsula. It was one of the slowest transcontinental flights in recent history, but it carried a precious billion-dollar hitchhiker that would put Americans back in space—if it survived its first orbital test flight.

On the morning of March 24, I joined a gaggle of television news and camera people beside the Kennedy Space Center's three-mile Shuttle runway. From the air, it looks

The Vehicle Assembly Building (VAB) is one of the world's largest buildings. It covers eight acres and has a volume of 129,428,000 cubic feet. It is 525 feet tall, 716 feet long, and 518 feet wide. It has more than 70 lifting devices, including 250-ton bridge cranes. This aerial view to the northwest shows the VAB and its environs. Behind the VAB is the 15,000-foot shuttle runway, which is 300 feet wide and has a 1,000-foot safety overrun at each end. It is aligned 330° north-northwest and 150° south-southeast. To the north stands a spare mobile launch tower.   (NASA)

like a segment of interstate highway they forgot to finish. Scene 3 was coming up.

It was a bright day, all blue and dazzling as the sun climbed out of the ocean. About one-half mile to the southeast, busloads of KSC workers were arriving at a roped-off reception center to greet the unfinished spaceship. They brought spouses and children, and the scene had the semieuphoric aspect of opening day at the circus.

I looked for the 747 with the bird on top to approach from the northeast, but unexpectedly there appeared in the south a fat pelican with a sea gull riding on its back. The two were in level flight about 1,500 feet above the Atlantic beaches and quickly turned out to be none other than NASA's second-hand 747 with *Columbia* perched on top.

Fitz Fulton and his crew had flown across South Brevard County on the route the returning orbiter might take on a southeast approach. They had turned north and were flying low over the surf with a deafening roar to the amazement of about 20,000 bathers who had never seen anything like that lash-up before.

It was a bit of grandstanding in the best circus tradition, and it did what it was calculated to do—it pro-

duced awe and excitement in the onlookers, many of whom were old enough to be paying for it.

Fulton circled the Space Center and then landed the jumbo jet with its 35-ton payload smoothly. I could hear the hiss of tires over the reverse thruster roar as the 747 rushed past in a gale of hot wind.

Columbia looked unfinished and battered, although it had lost only one temporary tile on the trip. It was carried grandly to the demate scaffold and a makeshift platform erected for the reception. The crew climbed down the ramp along with several passengers, including John W. Young, who would command the first orbital flight of the new spaceship.

Noted for the brevity of his remarks at public functions, Young responded succinctly to a newsman's question: When are you going to launch?

"As soon as Kappy [Walter J. Kapryan, launch director] says we're ready, and not before," replied Young. Having flown Gemini and Apollo and walked about on the Moon, he had plenty of experience dodging questions from the news media. This one was a good one to evade, because neither Young nor Kapryan, nor anyone else in the space agency, could answer it.

Orbiter 102, the *Columbia*, arrives at Kennedy Space Center, March 24, 1979. Crews worked through the night to uncouple the orbiter from its Boeing 747 carrier in the mate-demate device. (NASA)

Demated, *Columbia* is hauled by tractor March 25, 1979, to the Orbiter Processing Facility to be processed for its first flight. Gaps in the surface of the nose and fuselage show where thermal protection tiles are missing. (NASA)

Columbia enters the Orbiting Processing Facility March 25, 1979, to have installation of the heat shield completed. The slot at the top of the OPF door (upper right) allows the orbiter's vertical stabilizer to pass through. (NASA)

Panic

Why was *Columbia* moved from California to Florida before its manufacture was completed? The only explanation that fits is panic.

Since the fall of 1978, pressure from White House, congressional, and military sources had been mounting on NASA to get the Shuttle off the ground. There was an unmistakable political component to it, too, for the series of launch delays added fuel to the criticism of the Carter administration's failure to realize its goals.

Moreover, the launch postponements raised questions in Congress about management competence, since the Shuttle directorate was unable to predict a launch date that would hold. This was a new experience. Delays as persistent and as long as those in completing a Shuttle flight article had not occurred before. A year had been lost on Apollo in 1967 as a result of the cockpit fire in *Apollo 1,* but the flight schedule moved along rapidly after that. In the spring of 1979, the Shuttle was falling two years behind the 1971 schedule and delay was costing in excess of $1 million a day in working time.

The Shuttle directorate thought it was good strategy to bring *Columbia* to Florida. They believed they could complete the tiling, install the engines, finish the interior, and simulate flight tests while the vehicle was being processed for launch. In addition, the transfer created the illusion of forward movement in the program.

In actuality, the strategy didn't work and the program was slipping backward. Installing the heat shield turned out to be immeasurably more complex than expected, and moving the orbiter to Florida made the work more difficult.

Rockwell tilers and their families also had to be transferred from California to Florida. Part of the force became dissatisfied or proved inefficient and returned to California soon after it arrived in Florida. Local replacements had to be recruited and trained. Fortunately, college students became available as summer vacations started. They were great. They learned fast and most of them regarded the work as a quasi-public service, although at $6.50 an hour with lots of overtime, no one was making a sacrifice.

The Orbiter Processing Facility, where *Columbia*

roosted while its feathers were being glued on during the remainder of 1979, had not been designed for manufacturing. As many as 400 workers crowded into the hanger, sometimes overflowing the work spaces and standing in long, restless lines in the cafeteria nearby during the lunch break.

Many of them quickly resorted to "brown-bagging it" for lunch. They sat on the white concrete apron outside the hangar in what shade they could find, washing down sandwiches with soft drinks and weak coffee from a commissary van. Looking seaward, they could see the tower of Launch Pad 39-A three miles away, empty and waiting. It was a highly visible stimulus.

Proximity to the pad was imagined to inspire more of a gung-ho attitude among tile installers than would the industrial environment of the factory at Palmdale. So suggested Colonel Robert F. Overmyer, a Marine Corps test pilot who had joined the NASA corps of astronauts from the defunct Air Force Manned Orbital Laboratory program.

Overmyer served as deputy to Kenneth Kleinknecht, the Johnson Space Center's Shuttle manager. Although administratively part of flight operations at Houston, Kleinknecht and Overmyer worked much of the time at Kennedy, managing launch preparations. Their main concern was the heat shield. In fact, Kleinknecht, 60, had been recalled from NASA liaison duty in Paris to expedite tiling. He was one of the senior engineer executives who came to NASA when it absorbed the old National Advisory Committee on Aviation. They formed the cadre of the manned spaceflight engineering staff.

Overmyer seemed to be the ideal assistant to Kleinknecht. The energetic 38-year-old astronaut infused his own enthusiasm for the Shuttle into the tiling force. By late summer, the force had been increased to 1,400 men and women.

A morale problem appeared as summer ended. Families of the transplanted Californians wanted to go home so that their children could return to school in their own communities. Some were not happy with the Brevard County school system, where they said teachers complained of being underpaid and overtaxed with large classes.

By fall, it had become evident that tiling *Columbia* was a major pacing item that would determine the date of the launch. This was an unexpected development. Thermal protection on manned spacecraft had not been seen as a problem before, but now it threatened to delay orbital flight indefinitely.

The dimensions of the problem were not realized until testing showed that a large part of the tile shielding was unstable under stress. Even then, the weakness seemed to be localized as a bonding difficulty. But it was much more than that. The whole concept of applying the shielding like stucco on a building seemed to be faulty; the future of manned spaceflight by this country depended on correcting it.

A New Invention

The orbiter's thermal protection system is a new invention, a major one in the space age. Unless it works, the project of creating a reusable space vehicle falls apart. And there can be no economical program of manned spaceflight without a reusable vehicle.

The function of the shield is to allow the orbiter to reenter the atmosphere safely without burning up. It is to last 100 missions.

There are several kinds of heat shields. The system designed for the Shuttle is the most advanced—in theory. On the Mercury, Gemini, and Apollo capsules, the shield consisted of an epoxy-resin compound that removed heat by flaking off. This type of "ablative" shield can be used only once.

A new thermal protection material was developed by NASA's Ames Research Laboratory and the Lockheed Missiles & Space Company. It was light in weight, impervious to heat, and easy to manufacture. It consisted of pure silica fibers, made of sand and stiffened with clay. A slurry of the fibers mixed with water was cast into soft, porous blocks. Then a binder solution was added. When the stuff dried to a rigid block, it was cut and machined into six- and eight-inch square tiles, which were molded to fit the orbiter's contours.

Resistance of the tiles to temperature change is astonishing. One can hold a tile by the edge while the center glows red hot and not feel heat. The mixture can be pulled out of the kiln at 2,300° F. and plunged into cold water without damage.

There are about 24,000 of the six-inch-square tiles. They range from one to five inches in thickness. The thicker tiles are placed on areas where high frictional heat at reentry is expected. These six-inch tiles are coated with a mixture of tetrasilicide and borosilicate glass. It gives them a glossy black sheen. On the orbiter, they appear as the black part of the black-and-white pattern.

The white part of the pattern is formed by about 7,000 tiles eight inches square. They are thinner, ranging from one-half inch to three inches in thickness. They are designed for areas of lesser heating and have a white coating of silica compounds and aluminum oxide.

The 31,000 tiles cover most of the fuselage and wing. The nose cap and wing leading edges are protected by a super-heat-resistant material called carbon-carbon. It is made of laminated nylon cloth dipped in phenolic resin and alcohol and cured by high heat to convert the resin and alcohol into carbon. The material is then coated with a mixture of silica, silicon, and silicon carbide at high heat. The laminate that results is capable of resisting temperatures above 2,500° F.

Because the orbiter would reenter the atmosphere nose up, the thicker, black tiles are placed on the underbody, body flap, and lower wing surfaces, where heat up to 2,300° F. was anticipated. The thinner, white tiles shield the upper fuselage, top wing surface, and rudder, where temperatures were not expected to exceed 1,200° F.

The upper cargo bay doors and the pods containing the orbital manuevering engines, where heat is not expected to exceed 750° F., are shielded by a heat-resistant material called nomex felt. Ceramic tiles were added to the pods after the first orbital test flight in 1981.

Why It Took So Long

The process of affixing the tiles to the metal skin of the orbiter was—and is—incredibly complicated. By the end of 1979, when it seemed an endless task, NASA officials swore that if they could find any other shielding that would be reusable, they would resort to it instantly. But although headquarters authorized research into alternate thermal protection systems, the shield of sand remained the only one that seemed to be feasible.

When *Columbia* was ensconced in the Orbiter Processing Facility at the end of March 1979, it took one worker about 40 hours to install one tile. The process was speeded up to 1.8 tiles during the summer. That is, 1.8 tiles per worker per week!

One afternoon in May, I watched the tiling effort in the OPF. The Brobdingnagian spacecraft was surrounded by metal scaffolding on which Lilliputians stretched, stooped, squatted, or sat, with feet dangling over the side of the steel pipe scaffold.

Columbia was suspended from a beam by steel cables, its lower fuselage about ten feet off the floor, wheels down. A hefty young woman in coveralls, her long blond hair caught up in a white baker's cap, and a bearded young man in T-shirt, jeans, and sneakers worked side by side in preoccupied silence.

Each of the thermal protection tiles applied to *Columbia* was packaged individually and its position on the vehicle precisely charted. Here, Rockwell International technicians mount each tile separately.   (NASA)

Tile installation was hard work, demanding utmost care. Many of the installers were hired by Rockwell from Brevard County communities near the Kennedy Space Center.   (NASA)

A technician proof tests a tile to determine how tightly the tile adheres to the skin.   (NASA)

"College kids," said the Rockwell tile supervisor. "Pretty good, but we'll lose them when their classes start."

On the floor, beneath the 78-foot delta wing, three tilers in Rockwell coveralls were poring over a chart. The undersides of the vehicle were black with the high-temperature, six-inch square tiles, except for gaps where

tiles were yet to be fitted. Working in pairs, installers were ranged along the fuselage, some standing on low rungs of the scaffold, others working 20 to 30 feet above the concrete floor at the rear fuselage and the rudder. Several of the installers were standing, stretching and working with arms over their heads, pausing now and then to rest from this tiring position.

In contrast to the cavernous gloom of the Vehicle As-

Tile installers use a gauge to measure curvature of a part of *Columbia*'s upper wing surface in preparation for bonding tiles to the aluminum skin.   (NASA)

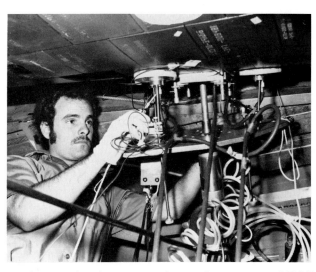

Proof testing the tiles was a 24-hour-a-day process.   (NASA)

Installers check the fit of a tile to its slot before bonding it to the orbiter.   (NASA)

sembly Building (VAB), the interior of the Processing Facility was brilliantly lit from above. Tiling required high illumination for precision work. It was not a job that could be done without training.

Each tile was marked to identify its position on the structure. The "ID" matched its assigned position on the aluminum skin. Each tile was packed separately in a plastic shipping container and protected from contamination by a thin plastic wrap.

The brittle square could not be touched by bare fingers for fear of leaving a thin smear of oil on the surface that would damage it in vacuum. All installers wore plastic gloves when wrapping or unwrapping or setting the tiles in place. Every step of the long process was recorded on a card dossier for each tile, and all 31,000 cards were filed in a fireproof steel cabinet.

When an incoming batch of tiles was delivered, usually by air freight from the factory, each tile was first inspected and then fitted temporarily into its slot to verify its dimension. Then it was labelled, placed in a foam-padded box for storage and shelved like a book in the library until the installer was ready for it. Its location was recorded in a computer. A printout showed where to find it.

These were the preliminary steps. The bonding pro-

cess began with the tile once more being fitted temporarily into place. Then shims were inserted around the edges. The shims were precision wedges that temporarily kept the tiles slightly apart. Each edge of each tile was separated from its neighbor by a gap of 25 to 60 mils. The gap would allow the tiles to be slightly displaced by expansion and contraction of the aluminum skin without clashing.

When the tile was fitted into place, a checking device called a "comparator" was applied to verify the fit. If a light appeared on this instrument, it indicated an irregularity—which could not be tolerated. One tile cannot be higher or lower than its neighbors, because that would cause turbulence in the air flow over the surface during atmospheric flight.

When no light appeared, the tile was deemed to have passed the fit check. It was bonded to a pad of nomex felt with a carefully measured amount of silicone rubber cement. The felt served as a "strain" pad, isolating the rigid, brittle silicate tile from the flexing of the aluminum walls.

It took about 16 hours for the cement to cure. Then a silicone rubber primer was applied to the metal skin where the tile, now bonded to the felt pad, was to be affixed. Next, an epoxy rubber cement was brushed over the area.

The tile was then set in place, felt pad down, and held firmly by a jack for another 16 hours. The shims were removed and the narrow gaps at the edges were filled with strips of nomex felt.

Where the Strategy Failed

There is a similarity of heat shield and Space Shuttle Main Engine (SSME) difficulties. The development strategy deliberately skimped in testing both. SSME components, as I have mentioned, were not tested adequately outside the engine. When they failed in engine tests, they damaged the engine.

The heat shield was not tested for mechanical integrity until it was 75 percent installed on *Columbia*. Then the failure of tiles to stick under pull stress presented NASA with a new technical problem toward the end of orbiter development. The problem, which should have been solved at the start, was purely mechanical. There was nothing wrong with the heat resistance qualities of the shield. They had been thoroughly proved in the laboratory. But many would not stay on and some would split under stress, problems no one had expected. Even after the tiles fell off during the test hop at Dryden, the full scope of the problem apparently was not realized, for NASA was pressing ahead with plans for a fourth-quarter 1979 launch. It was assumed that the 7,800 tiles that were missing when *Columbia* arrived at Kennedy in March could be installed by early fall.

This expectation went up in smoke when the tiles already on the vehicle were subjected to a series of pull tests in September. The tiles were pulled outward. It was not necessary to pull them off. Sounds produced by the pulling were carefully monitored; an analysis of their pitch and volume indicated whether the tile would stick under aerodynamic stresses during launch and reentry.

Thousands of tiles failed the tests and were replaced. The testing and replacement process continued through fall and winter.

The testing showed that many of the tiles were splitting horizontally under stress, rather than at the bond.

Why had NASA failed to run flight tests of this new shielding on aircraft or reentry vehicles earlier in the development program? I raised this question several times at status briefings. The answer was always the same: it would have been too costly.

The ablative epoxy-resin shielding used on the capsule spacecraft—Mercury, Gemini and Apollo—had been tested on nose cones repeatedly. Why could the shield and sand not have been tested in this way?

Hindsight shows that the heat shield debacle, which was partly responsible for delaying the completion of the shuttle from late 1979 to early 1981, was simply a matter of oversight. The basic miscalculation was in thinking of the shield as an add-on, external fitting, rather than as a structural element. A reusable shielding was not stable as an add-on; it should have been built into the vehicle's structure, critics said.

This realization dawned on NASA and the contractor too late for *Columbia*. Somehow, the shielding had to be fixed to last at least for one or two orbital flight tests. Two methods of dealing with the tile integrity problem were debated throughout the agency. One was to provide the crew with the means of repairing the shield in orbit if tiles came off during launch. In-flight repair called for a manned maneuvering unit (MMU) that would enable an astronaut to skitter about outside the orbiter on an extra-vehicular activity (EVA) operation. The astronaut would inspect the shielding before the decision to reenter was made and, if tiles were missing, would fill the gaps with a heat-resistant compound from a tile repair kit, which also would have to be developed.

The second method, advocated by the Johnson Space Center, was to strengthen the tiles to the point where there was no question about their integrity. One way of doing this appeared to be to make the base of each tile denser to avoid splitting. A compound had been tested that would "densify" the tiles, but applying it would require the installers at Kennedy to remove about one-third of tiles already bonded to the vehicle, treat them to make them stronger, and put them back on. The process would take months.

Developing a maneuvering unit had been considered long before the tile debacle. An MMU had been tested on *Skylab*. Such a propulsion device would enable crewmen to inspect satellites and perform work on space structures.

Three types of repair kits were studied at Marshall, with such gap-filling materials as silicon carbide mixtures or epoxy foam to replace missing tiles. As an alternative to an EVA, the crew could inspect the heat shield by means of a remotely controlled television camera mounted on an extendable boom attached to the cargo bay. It would save a trip outside if no damage was visible.

Although a precedent for repairing a vehicle in orbit had been established on *Skylab*, neither Young nor Crippen was eager to be committed to an in-flight repair on the first mission. Both would be preoccupied learning the ropes of piloting a brand new spaceship and constantly checking its systems. An EVA would seriously disrupt the busy schedule of the 54-hour first orbital flight.

The decision by headquarters was to omit an EVA

The cockpit of the Space Shuttle *Columbia* is a study in complexity. Commander occupies the left-hand seat, as in a conventional aircraft, and the pilot the right-hand seat.   (NASA)

As though taking a page from Buck Rogers–Superman fantasy, NASA contracted with Martin Marietta Aerospace to develop an astronaut propulsion system, the Manned Maneuvering Unit, or MMU. Attached to the space suit's portable life support system, the MMU would enable the astronaut to fly a limited distance from a spacecraft. A prototype of an MMU that would enable a person to realize an age-old dream of flying Peter Pan style was tested in 1973 by Jack R. Lousma on *Skylab 3* in the forward dome of *Skylab*'s huge orbital workshop. Propulsion was provided by 24 nitrogen gas jets, which the astronaut operated by means of hand controls.   (NASA)

them with ludox, a silica-boron compound. At first, about 4,500 tiles were to pulled off, densified, and replaced, but the number steadily grew once the process started.

Second Installment, 1979

Meanwhile, Orbiter 101, the *Enterprise*, returned to public view once again at KSC. Fitz Fulton and his crew flew it on the 747 to Kennedy from Marshall on April 10. Since there were no tiles to worry about, the trip was smooth; Fulton made three touch-and-go passes before landing on the three-mile runway. On each pass, the 747 with the orbiter on top came down until the wheels touched and then, in a demonstration of power, the pilot lifted the big airplane and its cargo to soar around the space center and come in again. The maneuvers delighted the crowd of about 400 space workers and their families who were on hand to greet the arrival of 101.

The *Enterprise* was mated to the external tank and SRBs in the Vehicle Assembly Building as a test of the assembly process. On May 1, this assemblage was transported on the giant crawler to pad 39-A, where it was erected to test pad operating procedure. The crawler, a tractor

commitment on the first orbital test, densify critical tiles, and fly a "benign" mission so as to keep stress on the heat shield to a minimum.

The density of the tiles was to be increased by treating

Enterprise emerges from the Vehicle Assembly Building at Kennedy Space Center May 1, 1979, aboard the giant tractor-transporter on the 3.5-mile journey to Launch Pad 39-A.
(author photo)

A rooftop view of *Enterprise* being moved out of the Vehicle Assembly Building mated to a dummy external tank and dummy solid rocket boosters. (NASA)

Enterprise reaches the launch pad after the slow, nine-hour transfer from the Vehicle Assembly Building. It remained on the pad five weeks for fit and function checks, breaking trail for the arrival of *Columbia*. (NASA)

with a platform as big as a baseball diamond, had not been used since the summer of 1975 when it carried an Apollo-Saturn IB to the pad for the launch on the joint flight with the Russian Soyuz.

The fully assembled Shuttle remained on the pad until May 23, when it was taken down, hauled back to the VAB, and disassembled. Orbiter 101 was then flown back to California to be used for exhibitions and ultimately spare parts.

It was a brief distraction in the long pull to get *Columbia* ready for flight. After the Saturn 5–Apollo, however, the Shuttle, looking squat and very bulky, was not a romantic sight on the pad. It had the unglamorous, businesslike aura of a streetcar.

In Washington, seemingly endless heat shield and engine delays were generating concern, especially about the rising cost of the program. The climax came May 1, when Administrator Frosch appeared before the Senate Subcommittee on Science, Technology, and Space with a request for a supplement of $220 million to the funds budgeted for the Shuttle in the 1980 fiscal year.

It was the second supplement since March, when Yardley had pleaded for a $185-million addition to the 1979 fiscal year budget. At that time, the Shuttle boss had pledged that no further supplementary funds would be required.* But two months later, the administrator himself was asking for another transfusion.

The subcommittee chairman, Adlai Stevenson, was dismayed. "Dr. Frosch, two months ago you assured us that the $185 million for fiscal 1979 and the budget you requested for fiscal year 1980 would be adequate barring any serious development problems," he protested.[1]

Frosch allowed there had been unanticipated serious development problems. Senator Howard W. Cannon (D-Mo.), chairman of the full Committee on Commerce, Science, and Transportation, wondered why they had not been anticipated. In a letter to Frosch, Cannon suggested that the new request "raises a basic question with respect to the adequacy of the Shuttle management system."

Headquarters reacted to the heat by organizing an inspection task force to assess administrative efficiency at the space centers. It was merely a gesture, futile and superficial. Management of the program had miscalculated its order of difficulty because of the assumption that building the Shuttle would not require new technology.

Moreover, the Office of Management and Budget (OMB) had persistently cut Shuttle appropriations, forcing the space agency to defer work from one year to the next and delay the "production" orbiters. A fleet of five orbiters had been planned, but one was eliminated for economy. That left four: Orbiter 102, the *Columbia;* Orbiter 099, the *Challenger;* Orbiter 103, the *Discovery,* and Orbiter 104, the *Atlantis.* All were named after fa-

mous research ships. As distinguished from the development orbiters, 102 and 101, the production orbiters, 099, 103, and 104, would be operational from the start.†

Because of OMB cuts, NASA was forced to transfer construction funds for Orbiters 099, 103, and 104 to 102, in order to get it off the ground as early as possible. The executive budget agency lopped off $85 million from NASA's fiscal 1974 request, $89 million from the fiscal 1975 request, $45 million from the 1976 request, and $100 million from the 1977 budget request.

Fund transfers and work deferrals to which the agency resorted as a result of the cuts now had to be made up— in addition to escalating engine and heat shield development costs that had not been anticipated.

Reallocation of money from the future orbiter fleet to *Columbia* amounted to $100 million in the 1978 fiscal year and $70 million in fiscal 1979. Added to these sums were $25 million NASA received from the Economic Stimulus Appropriation Act and $30 million left over from the 1975 joint flight with the Russians.

When all of these sums are added to the $185 million supplement to the 1979 budget and the $220 million supplement to the 1980 budget, the excess funds for developing *Columbia* amount to $640 million. And that was not all.

Before *Columbia* could clear the OPF to start its journey to the launch pad, Frosch would again come to Congress for a third supplement—$300 million, of which he would be granted $285 million. Thus, the additional funds required for *Columbia* reached $925 million by the end of 1979, over and above those actually budgeted in the 1979 and 1980 fiscal years. Delay in the construction of at least two production orbiters that had been set back by the transfer of funds to *Columbia* was not made up by the supplements. Orbiter 099, being modified to replace the *Enterprise,* had been due for delivery at Kennedy Space Center in February 1981; it was delayed initially to March 1982 and did not arrive until July. Orbiter 103, originally scheduled for delivery at Vandenberg Air Force Base, California, in March 1982

---

*Yardley had qualified this with the condition that unforeseen problems would not occur. But they did occur, as they had all along.

†Orbiter 099 originally had been planned as a structural test article but was redeveloped as a flight article after Orbiter 101 was given the test role because of overweight.

was not ready for delivery until autumn 1983, and then it was destined to go to KSC because Vandenberg was not ready. Vandenberg was being developed in 1983 as the Shuttle's West Coast launch and landing site for polar orbit flights, under Air Force control for the most part. Consequently, there was Department of Defense concern, transmitted to Congress, about the inordinate Shuttle delays. The DOD was expected to use the Shuttle to boost heavy military satellites in the 1983–1985 time slot.

As a result of the headquarters "efficiency" investigation at the Centers, some minor reshuffling was ordered. At Kennedy, Walter J. Kapryan, one of the most capable and experienced launch operations directors in the agency, retired as Shuttle Operations Director after working in government aerospace research agencies for 32 years. The Kennedy Space Center director, Lee R. Scherer, resigned rather than accept what seemed to be a promotion to headquarters in Washington as associate administrator for external relations, a public relations function. Scherer was succeeded at Kennedy by Richard Smith, who had handled the *Skylab* plunge with such aplomb and effectiveness.

It would have been easy for Frosch to launch a witch-hunt or find a scapegoat for the Shuttle problems, but he kept the "efficiency" review low key. Later in the year, he laid out the whole problem to his boss, President Carter, who understood that he had inherited a technological miscalculation from Nixon and Ford.

Because the latent military importance of the Shuttle was now being perceived as an urgent matter, Carter assured Frosch of unstinting support. The Shuttle had finally evolved into a national necessity.

The Welding Wire Mistake

Summer 1979 seemed to be the nadir of the Shuttle program. The heat shield looked like a disaster and the engines were failing.

Frosch told the Senate subcommittee that "work on *Columbia* at the Kennedy Space Center is proceeding slowly."[1] The tile problem was "most severe," he said. There were still 8,000 tiles to be installed, about 200

more than had remained to be fitted the previous March. Tiles showing defects were pulled off faster than replacements were put on. Like Alice in Wonderland, NASA had to run as fast as it could just to stay in one place. It was losing. By the first week in June, an installation rate of only 200 tiles a week had been achieved, compared to a planned rate of 650.

At the National Space Technology Laboratory in Mississippi, efforts to fire the Main Propulsion Test Article (MPTA), postponed from May 15, continued into the fall. On October 24, a third attempt was scrubbed when a hydrogen leak was found in a feed line. As the delay stretched to six months, the test was rescheduled for November 4.

This time, the three engines were ignited, but at nine seconds, they cut off. A sensor on engine 3 reacted to overpressure in the high-pressure liquid oxygen pump and shut down the whole propulsion system.

This relatively minor event had major consequences. The abrupt shutdown of the three engines caused the hydrogen line in engine 1 to rupture. Hydrogen gushed out, erupted in flames. Engine and test stand were damaged.

The investigation showed that the part of the hydrogen line that failed was the steer horn. It was the same failure that had occurred on May 14 on a single-engine test at Santa Susana—the failure that had persuaded the test crew in Mississippi to postpone the May 15 MPTA run until they could check out the cause.

When the steer horn failure had been determined at Santa Susana, the engines in Mississippi had been inspected for signs of the same weakness. They seemed to be okay. They were not changed. After six months of delay, the steer horn had failed again.

On November 19, NASA headquarters announced that the November 4 steer horn failure could be blamed on the use of improper welding wire, which had weakened the line. Instead of Inconel 718, a thinner weld wire, Inconel 600, had been used. The mistake was blamed on a supplier. Now the agency was confronted with the necessity of inspecting every engine in its inventory, including the three that had been installed in *Columbia* during the summer, engines 2005, 2006, and 2007.

By the end of 1979, persisting uncertainty about the tiles and the reworking of the engines caused the launch date to slip into the summer or early fall of 1980 after a plan to try for a March 30 launch date was shot down. Technically, it was possible to achieve the March 30 date by rushing completion of the shield and engine reworking, but there was great risk. Both the shielding and the engines, as they stood then, were undependable.

Cavalry to the Rescue

Both House and Senate space committees were increasingly critical of Shuttle delays. Had a Russian Shuttle suddenly materialized, there would have been a congressional investigation of the NASA program. At times this seemed to be imminent, but Congress was too busy with more pressing matters to pursue it.

The House and Senate space committee hearings in mid-1979 and later brought out the expectations of the Defense Department for using the Shuttle for military missions. Although NASA had earlier soft-pedaled the Shuttle's military importance, there was no doubt about its utility as an adjunct to the national defense armamentarium.

During hearings before the House Subcommittee on Space Science and Applications at the end of June, Frosch referred to the military significance of the Shuttle in defending his request for the $220-million supplement. Failure of Congress to provide these extra funds would delay the production of Orbiters 103 and 104, which the Air Force had planned to use as soon as it could get them.

The subcommittee invited Dr. William J. Perry, Undersecretary of Defense for Research and Engineering, to discuss DOD requirements for the Shuttle, a subject hitherto not emphasized in Shuttle hearings by space committees. "The Defense Department is becoming increasingly dependent, first of all, on space and as time goes by, on specifically the Shuttle," he said. "Our dependence on space for navigation, communications, early warning, surveillance, and weather forecasting is be-

coming increasingly greater in the decade of the '80s.

"We are beginning in 1982 to transition to the Shuttle vehicle, and ultimately we will be conducting all of our space operations with the Shuttle."[2]

Perry emphasized that DOD planned to use the shuttle as soon as it could get it. In 10 years, DOD projected 113 shuttle launches, and by the end of that time, Defense would be launching the Shuttle 15 times a year, he said.

Perry left no doubt that DOD intended to replace all of its expandable launchers with the Shuttle. By the middle 1980s, he said, it would be totally dependent on the Shuttle for satellite launches.

Other military arrangements for the Shuttle were disclosed at a meeting of the 17th Space Congress of the Canaveral Council of Technical Societies, at Cocoa Beach, Florida, in May 1980 by Maj. Gen. James H. Marshall, commander of the Space and Missile Test Organization at Vandenberg. For security reasons, Marshall said, the DOD planned to operate military Shuttle missions separately from NASA. They would be controlled from a Consolidated Space Operations Center that DOD was building at Colorado Springs at a cost of $500 million. Only NASA flights would continue to be controlled from the Johnson Space Center at Houston.

Marshall said that DOD had ordered Titan rocket engines to be strapped on to the Shuttle external tank for additional propulsion. This was necessary to lift 16-ton payloads, which Defense planned to launch from Vandenberg. *Columbia* and its three sister orbiters were limited to 12 tons of payload on polar launches. On eastward launches from Kennedy, taking advantage of the Earth's rotation, the Shuttles could lift a heavier payload, but they could not attain the load maximum of 32.5 tons, or 65,000 pounds, they were originally designed to lift without extra propulsion.

Although considerably degraded from the original conception of 1969, the Shuttle was approaching reality at the end of 1979. In ten years, the United States had created a transportation system that could be considered the nucleus of a U.S. Space Force. Its commercial and military capabilities were open ended.

During the first quarter of 1980, although engine and heat shield problems persisted, their solution finally came into view. The years of frustration were coming to an end. Still, it would take another year before *Columbia* would be rolled out of the Orbiter Processing Facility, mated to the external tank and solid rocket boosters, and mounted on the crawler to be trundled to the launch pad.

The Universe or Nothing

Since the time of Thomas Malthus (1766–1834), warnings that mankind exists on the thin edge of ecological disaster have gone unheeded—perhaps because doomsday is constantly being invalidated, or at least deferred, by advances in technology.

Malthus held that the human population increased faster than the food supply unless checked by famine, disease, and war. In his view, poverty was the inescapable result of the variance in growth whereby the human population increased exponentially and its sustenance in linear fashion.

This lethal equation was invalidated by the Industrial Revolution and its "spin-off" in the underdeveloped countries, the Green Revolution. However, in 200 years, industrialization generated its own doomsday equation by using up the nonrenewable resources of materials and energy on this planet.

Neo-Malthusians perceive this depletion as the process that will halt the growth of industrial civilization and world population in a century or less. They say it will force the people of the Earth not only to reduce their numbers but also to regress to a simpler, more bucolic life-style.

But—just as the rise of technology in the eighteenth and nineteenth centuries negated or postponed the Malthusian day of reckoning, so does the rise of space technology in this century offer a way out of the dilemma of resource depletion. It does this, space planners say, by providing mankind with access to the vast resources of the Solar System. That is the long-range purpose of the Space Shuttle Transportation System, in their view. It extends man's reach for the necessities of life.

One of the most influential neo-Malthusian documents since Malthus' *Essay on the Principle of Population* (1798) is *Limits to Growth*, a report issued by a Massachusetts Institute of Technology research team for the

Club of Rome in 1972.* It warned that if present growth trends in world population, industrialization, pollution, food production, and resource depletion continue unchanged, the limits to growth on this planet will be reached some time in the next 100 years. The most probable result, the report said, will be a rather sudden and uncontrollable decline in both population and industrial capacity.

The alternative to this fate is to alter these growth trends, the report said, and to establish a condition of ecological and economic stability that is sustainable far into the future. How far the report did not say. Unless mankind reverted to the stone age in population and technology, it would inevitably use the nonrenewable resources (fuel, metals, etc.) it needs for survival to the point of unavailability or utter exhaustion.

The report envisioned a state of global equilibrium so designed that the basic material needs of each person on Earth are satisfied. How this utopian condition is to be realized in a fiercely competitive world was not spelled out, but the report warned that the sooner we start working toward it, the better our chance of success.

In a time of burgeoning ecology consciousness and rising concern about the environment, *Limits to Growth* has become the Magna Charta of the ecology movement. A *New York Times* reviewer hailed it as "one of the most important documents of our time." Its central thesis was that the Earth is a finite habitat and that mankind is going to wake up some morning in the next century to find that it has run out of stuff.

Since 1650, the report said, the human population has been growing exponentially at an increasing rate. Starting at 500 million then, it doubled in 250 years.† In 1970, the population was 3.6 billion, with a doubling time of 33 years.

This expanding population consumes resources at an accelerating rate. If it continues, according to the report, the available world supply of aluminum will be depleted in 31 years, chromium in 95 years, coal in 111 years, cobalt in 60 years, copper in 21 years, gold in only 9 years, iron in 93 years, lead in 21 years, natural gas in 22 years, and petroleum in 20 years.

Although economists and mineralogists have disagreed with these estimates, the message can hardly be disputed. Sooner or later, the nonrenewable resources that provide us with food, shelter, and energy will be gone. Then what?

Shortly after *Limits to Growth* appeared, its warning was dramatized by the oil shortages of 1973. Although a managed effect by Middle East oil-producing countries, the dry spell told us something we knew but didn't quite believe: fossil fuels are limited.

In a more recent discussion of this dilemma, a book entitled *Entropy,* by Jeremy Rifkin and Ted Howard depicts the exhaustion of nonrenewable resources as a cosmic effect.[1] The effect is "entropy," the second law of thermodynamics, which states that matter and energy can be changed only in one direction—from the usable to the unusable; from the available to the unavailable. Entropy is what makes nonrenewable resources nonrenewable, at least on the human time scale.

In the preindustrial age, when people relied on the energy stored in plants, the carrying capacity of the Earth was about 1 billion people. With the transition to coal, oil, and gas, a higher energy flow was released from stored, nonrenewable sources. It supported an additional 2.5 billion people. When those resources are gone, the additional population cannot be sustained, according to Rifkin. As a consequence, he maintains, the Earth's human population must regress eventually toward a "sustainable solar age population" of about 1 billion people. Rifkin's "solar age" is the preindustrial age.

Like Malthus, the modern prophets of doom seem to consider humanity forever dependent on the resources of a single planet, the Earth.

---

*The Club of Rome was an informal group of about 75 scientists, educators, economists, industrialists, and intellectuals who began meeting in Rome in 1968 to investigate the future of mankind. *Limits to Growth* is a report by Donella H. and Dennis L. Meadows, Jorgen Randers, and William Behrens III for the club's project on the predicament of mankind. The report was published by Potomac Associates, a tax-exempt organization dedicated to research and analysis in Washington, D.C.

†The source of this figure is not clear. One wonders who was counting then.

But space technologists and theoreticians have an alternative scenario, with a more dynamic outcome. It envisions a Shuttle-based space transportation system capable of setting up mining camps on the Moon and moving nickel-iron–rich asteroids into low Earth orbit as a source of heavy metals, as well as constructing solar satellites and establishing colonies on the Moon, on Mars, and in wheel-shaped habitats in orbits equidistant from the Earth and the Moon.

The obverse of the limits-to-growth view of the Club of Rome is the expansionist view of what might be called the "Club of Space." Its technical solution to the planet's limitations rests ultimately on a reusable, cost-effective space transport. In this context, the Shuttle assumes evolutionary importance to the future of civilized societies.

Beyond the Limits to Growth

The neo-Malthusians and the Spacers perceive the Earth in different ways. The former perceive it as a prison, in which humanity is forever confined. The latter view it in terms of Tsiolkovsky's dictum that while Earth is the cradle of mankind, one cannot remain forever in the cradle.

Appearing before the Subcommittee on Space Science and Applications of the House Committee on Science and Technology January 25, 1978, Gerard K. O'Neill, professor of physics at Princeton University, advocated the space solution to the limits to growth. The fatalism of the limits, he said, is reasonable only if one ignores all resources beyond our atmosphere, "resources thousands of times greater than we could ever obtain from our beleaguered Earth."

O'Neill is the principal architect of the space colony "wheel," a huge satellite a mile in diameter.* He spoke in support of House Concurrent Resolution 451, which

---

*The feasibility of a space satellite colony was examined in a study at NASA's Ames Research Center, California, in the summer of 1975. The design called for a wheel-shaped structure rotating around its hub at one rotation a minute. Centrifugal force would simulate gravity for about 10,000 people living in the rim, where housing, shops, schools, and light industry would be located. The colony would feed itself by closed-loop agriculture, get its power from the Sun, and manufacture other colonies and solar power satellites.

This artist's concept of a space settlement for a population of 10,000 at a point in the Moon's orbit where gravitational forces are in balance was developed in a 1977 design study of space settlements at NASA's Ames Research Center. The spherical habitat, a mile wide at its equator, rotates and provides nearly 1 G of force simulating gravity. A small river would flow at the equator, with banks of lunar sand. A corridor along the sphere's axis of rotation would provide passage from the center to the poles, where the artificial gravitation would be reduced to near zero for recreation and microgravity experiments. The habitat would achieve some degree of self-sufficiency by growing its own food and recycling its water. Its economy would be based on manufacturing products such as electronic components, special alloys, and purified pharmaceuticals that could be made of higher quality in microgravity than on Earth. It would be an orbital city trading with Earth.   (NASA)

states: "This tiny Earth is not humanity's prison, is not a closed and dwindling resource, but is in fact only part of a vast system, rich in opportunities, a high frontier which irresistibly beckons and challenges the American genius."

O'Neill cited the alternative to the "limits" in terms of "fundamental facts of science that will never change." "First," he said, "while we search desperately for new energy resources here on Earth, a few thousand miles above our heads there streams by constantly, night and day, a flood of high-intensity solar energy far greater than we would ever need."[2]

It does seem curious that the doomsday prophets overlook the energy potential of the Sun as the way out of the terrestrial fuel depletion problem. Each square meter of the Earth's surface that is open to the Sun at high noon receives the equivalent of 1 kilowatt of energy. The total (1.49×10^{22} joules) is 100,000 times more than the power produced in all the world's electrical generating plants, according to Peter E. Glaser, the engineer who invented the solar power satellite (SPS).[3]

At one astronomical unit, the distance from the Earth to the Sun, the collection potential of an area the size of the Earth equals in one day one-tenth of all the energy stored in fossil fuels, or eight trillion metric tons of coal, according to economist Klaus P. Heiss.[4]

Efforts to use solar energy on a small scale to operate pumps and to heat water were made during the nineteenth and early twentieth centuries. Solar hot water heaters appeared in Miami, Florida, in the 1920s and 1930s, and optical devices that focused sunlight to heat fluids that would drive turbines were developed in the first half of the century. But it was not until the early 1970s, when the economics of fossil fuel depletion and the hazards of nuclear power were realized, that solar energy became of major importance as an alternative.

Ground collection of sunlight is impeded by the day-night cycle, clouds, and dust and pollution in the atmosphere. The most efficient way of harnessing solar power for use on Earth is to collect it in space. The most direct method of doing this is establishing large arrays of photovoltaic cells in geostationary (geosynchronous) orbit. They would receive about ten times as much incident solar energy as a ground station would in the southern latitudes of the United States, according to Glaser. Unaffected by pollution, their lifetime is estimated at about 30 years. However, micrometeoroids (space dust) and cosmic rays would degrade the solar cells, which are made of silicon or gallium aluminum arsenide. Also, high-

Peter E. Glaser, internationally recognized as the inventor of the solar power satellite. He received a U.S. patent for it in 1973.   (Arthur D. Little, Inc.)

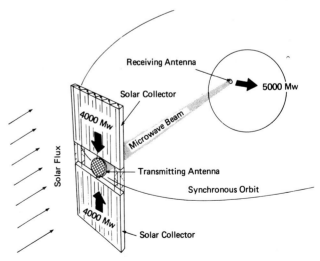

Design principles for a satellite solar power station are illustrated here. Solar electric energy collected by the photovoltaic cells is transmitted to Earth as microwave energy and transformed by the ground collecting station to electricity for distribution.   (Arthur D. Little, Inc.)

energy cosmic ray particles could damage satellite electronics. Periodic servicing would be required.

A number of solar power options have been studied by groups of engineers and scientists under NASA contracts. The one most widely supported is Glaser's concept of an SPS.

Glaser proposed an SPS capable of delivering 5 million kilowatts of electricity to a ground receiving station. The array of solar cells would be 10.6 kilometers long and 5.2 kilometers wide (6.6 by 3.2 miles). It would weigh about 2.2 million pounds.[5]

Electricity converted from sunlight by this array would be changed into microwave energy at common industrial frequencies of 2.4 to 2.5 gigahertz by crossfield amplifiers and linear beam devices. An antenna 1 kilometer (0.6 miles) square would transmit the beam to Earth. The receiving antenna would be located on a site 10 by 12

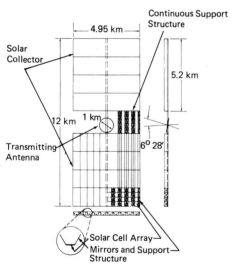

The solar power satellite structure shown here is 3 miles (4.95 km) wide and 7.45 miles (12 km) long. Mounted on a continuous support structure are two slightly offset rectangular collecting areas of photovoltaic cells. Each collecting area is 3.2 miles (5.2 km) long and 3.08 miles (4.95 km) wide. They are separated by a gap of 0.62 miles (1 km), in which the transmitting antenna is located. A transmitting antenna 0.62 miles (1 km) in diameter would require a receiving antenna 4.3 miles (7 km) in diameter on Earth. Mirrors at the end of the satellite structure would increase power generation.

(Arthur D. Little, Inc.)

kilometers (6.2 by 7.4 miles) in area, on land or off shore.

Power input from the satellite would be continuous, night and day, winter and summer. There would be no pollution, no waste heat would be dumped into the environment. Environmentalists have voiced concern about the possible health hazard to people exposed to the microwave beam. The pathological effect of such exposure is speculative, and somewhat controversial.

The American Institute of Aeronautics and Astronautics has found Glaser's model to be technically feasible. But "a considerable effort will be needed to determine its economic, environmental, societal, and political vitality," the AIAA commented.[6]

The economics of the satellite system have been examined by several economists. Heiss of ECON, Inc., a consultant for NASA, cited estimates showing that it would cost less to obtain large additional amounts of power from satellites than to obtain them from conventional generators.

In 1975, the total United States investment in electrical energy production, including plant and equipment, amounted to $200 billion, according to an Edison Electric Institute report, Heiss said. The system produced 17 quadrillion British thermal units, or, more concisely, 17 Quads of energy.

Heiss cited another estimate by the American Gas Association that the cost of adding 10 more Quads to the system would be $475 billion with conventional generating equipment. Alternatively, the cost of producing 17 additional Quads with SPSs in the 5 gigawatt range was estimated at $500 billion by four research and development companies, he said. On the basis of this estimate, the cost per additional Quad would be $29.4 billion with the SPS system, compared with $47.6 billion per Quad with conventional generating additions. The companies cited by Heiss were his own ECON, Inc.; Arthur D. Little, Inc.; the Grumman Aerospace Corporation; and Raytheon.[7]

Glaser believed that Shuttle technology would be able to meet SPS development needs for about a decade. Establishment of a global SPS system would require a heavy lift vehicle that could boost loads of 200 to 500 metric tons into low Earth orbit. Powerful upper-stage boosters or a space tug would be necessary in both cases to lift the satellite, presumably in sections to be assembled, to a 22,300-mile geostationary altitude.

Beyond the Shuttle

As I have related earlier, the Shuttle system was not conceived of as a single vehicle, although that is the way it has been developed because of cost constraints. The Shuttle concept included the space tug, but that was not funded. Consequently, interim or inertial upper-stage (IUS) rockets and payload assist modules (PAMs) would be provided as "interim" substitutes to ride along with payloads in the orbiter cargo bay to low Earth orbit, where they would be deployed to boost their payloads to higher orbits.

Even the Shuttle and the tug do not complete the transportation system as proposed in 1969. Space construction projects of the magnitude of the SPS, orbital factories, and space habitats require an interorbital transport, which might use nuclear propulsion. The system might also include permanent space station modules; a planetary lander; and the heavy lift vehicle. When all of these elements are considered, it is clear that the Shuttle is only the beginning of a transportation system that will enable humankind to exploit the "high frontier."

Two methods of building the SPS system have been analyzed. One uses terrestrial materials that are ferried to fabrication sites in low Earth orbit for shaping and then transferred to geostationary orbit for final assembly and activation. The other would use lunar aluminum, titanium, silicon, and glass for the fabrication of satellite components, including solar cell arrays and support structures at orbital factories.

It bears repeating, I think, that the second option presupposes an industrial infrastructure on the Moon, consisting of a permanent mining and processing camp. Routine Earth-to-Moon transport is presupposed, as is a geological survey of the Moon that might take decades if operated on the scale of the U.S. Antarctic Research Program. Lunar mining and processing represent a for-

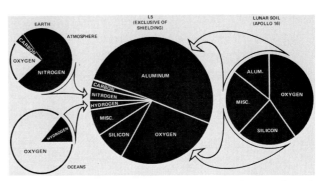

Because of lower transportation costs, structural materials such as aluminum, titanium, and glass would be obtained from the Moon and processed there or in an orbital factory. This chart shows which resources could be supplied from the Moon and which would have to be brought from Earth.   (NASA)

midable undertaking, but several studies suggest that it would be cheaper in terms of energy cost to build an SPS system with lunar materials than with materials from Earth. The cost of lifting material to geostationary orbit from the Moon is 4.5 percent of that of lifting it from Earth. Since energy is the main cost of developing the SPS the lunar option seems attractive.

In a report to NASA, the Space Systems Laboratory of the Massachusetts Institute of Technology developed this scenario of how the SPS could be built from lunar material: Aluminum would be obtained from lunar anorthosite, a common surface rock. Silicon, silicates, and titanium are available. The material would be processed at a lunar resource complex, hurled off the moon's surface by electric catapults to a cargo assembly point in space and, after being "packaged" there, ferried by tug to a space manufacturing facility in geostationary orbit.

Sections of the SPS would be fabricated at the facility and then lifted to geostationary orbit to be assembled.

The prospect of lunar resource development might be constrained by the United Nations "Moon Treaty," which calls for an indefinite moratorium on the commercial exploitation of extraterrestrial resources until an international authority is created to regulate it. This "Agreement Governing the Activities of States on the Moon and Other Celestial Bodies" was proposed to the United Nations by the Soviet Union in 1971, as Apollo astronauts were exploring the lunar surface. It was clear by then that the

This advertisement by United Technologies, one of America's leading aerospace manufacturing firms, appeared February 13, 1980, in four Connecticut newspapers.   (United Technologies)

Moon had abundant resources of aluminum and titanium.

At first, NASA supported the proposal. But as the development of the Shuttle lent lunar resources development some semblance of feasibility, the United States cooled off on the Moon Treaty. American aerospace executives began to perceive it as a threat to American economic interests on the Moon. Throughout history, the rule for exploiting natural resources of new lands had been first come, first served. Why change it? aerospace executives asked.*

Whether constructed of terrestrial or lunar raw materials, the economics of an SPS system are speculative.

*United Technologies Corporation denounced the Moon Treaty in February 1980 in full-page ads as shown above.

Some economists suggest that a $500-billion investment in it is not unrealistic. As the cost of fossil fuel rises, the SPS system may become competitive with terrestrial energy sources long before they are exhausted.

Civilizations of the past have mobilized resources for the construction of large-scale religious and military structures. The pyramids of Egypt and the Great Wall of China took centuries to build. In terms of social effort, they may be compared with the building of an SPS system by modern society, according to economist Heiss. Each probably used more than the 10 million man-hours represented by the $500-billion cost of a solar energy base in orbit.

Near Term: 1982–1990

None of these far-out scenarios is likely to be realized before the end of the century. Except for Spacelab missions, the Shuttle's principal function during the 1980s, as planned in the 1970s, was to demonstrate its cost-effectiveness as an alternative to expendable rockets. That was the initial justification for building it. Its role as the Conestoga wagon of the high frontier did not materialize until later.

Its first major logistical mission was to place the *Space Telescope* (ST) in a 310-mile orbit in January 1984.* This is the first of a series of national observatories in Earth orbit that are planned for the next century.

The telescope was designed to fit into the orbiter's cargo bay. It is 43 feet long and 14 feet in diameter and weighs 20,000 pounds. Perkin-Elmer built the optical telescope assembly, with an 8-foot primary diffraction mirror, for $58.5 million. The telescope also will carry a photometer, spectrometers, and two cameras, one of which is provided by the European Space Agency. The support structure was built by the Lockheed Missiles & Space Company for $72.8 million. It is to be serviced from time to time by the Shuttle.

The managers at the Marshall Space Flight Center expect the telescope to provide ten times better resolution

*Delayed to 1986 by telescope integration problems.

that it is possible to obtain with any instrument on the ground. It will extend man's "seeing" ten times deeper into space and enable observers to explore a volume of space up to a thousand times greater than they can see from the ground.

Of particular interest to searchers for extraterrestrial civilizations is the ability of the telescope to detect planetary systems around stars within 30 light-years of the Earth—where there are at least ten stars similar to our Sun.

A second major shuttle mission scheduled for the mid-1980s is *Galileo,* a new reconnaissance of Jupiter. A Jupiter orbiter and atmosphere probe are to be carried by *Challenger* into low Earth orbit and boosted from there by attached Centaur rocket to the big planet. Also in that time period, *Challenger* is schedule to lift and deploy a European Space Agency satellite into orbit for a look at the north pole of the Sun. The spacecraft is to be boosted on the solar-polar flight by a Centaur upper stage all the way to Jupiter, whose gravitation will sling the vehicle out of the ecliptic plane and on a flight path that will pass the solar pole.

Why Make It in Space?

Until 1981, the experience of manufacturing products in space had been limited to inconclusive tests of materials aboard Apollo, the Russian Salyut and Soyuz, and our *Skylab.* Speculation runs that certain high-technology parts and products can be manufactured with more efficiency and higher quality in space, because of the low gravity, than on the ground. Spacelab makes it possible to continue this experimentation. A group of West German firms has reserved one entire Spacelab mission for materials-processing experiments.

A deluge of cold water was thrown on space processing prospects in 1978 by the National Research Council's Space Applications Board on the basis of a study. Prospects were limited, the board said, and the NASA program suffered from poorly conceived and designed experiments and crude apparatus—from which it added, weak conclusions were drawn and overpublicized.

Although it conceded that meaningful science and technology can be developed from such experiments in space, the committee did not find an example of a process that would be economically justifiable. Even though low gravity appears to offer "certain capabilities" for materials studies, the board said, the instances where it is likely to be important for materials processing "will be few and specific." When gravity has an adverse effect on a process, stratagems for dealing with it can usually be found on Earth that are much easier and less expensive than recourse to spaceflight, the board said. It concluded that in sophistication and reliability, as well as cost, experimentation on the ground is generally superior to what can be expected in space.

The pessimistic view of this prestigious board does not seem to have discouraged anyone. At the end of 1979, at least 40 materials-processing experiments either were being planned or had been booked for Spacelab flights. Many of them were designed by West European industries.

Space processing advocates maintain that low-gravity experiments have demonstrated the advantage of the space environment as a manufacturing site for certain products. One major advantage is that materials can be levitated, that is, suspended in midair, without a container. Only small forces, such as electromagnetic, electrostatic, or acoustic fields, would be needed to control the position of large amounts of materials, according to Donald M. Waltz of the TRW Space Vehicles Division.[8]

Containerless processing eliminates impurities from the container and allows greater flexibility in heating and melting solids, Waltz said. He recalled that crystal-growth experiments on Skylab and the Apollo-Soyuz joint flight demonstrated distinct improvement in monocrystalline semiconductors. Their electrical properties were enhanced by higher purity, homogeneity, and structural perfection.

Waltz pointed out that these materials form the basis for all integrated circuit technology. The future market for them will be enormous, he said. Space factories could also produce glass of high purity and optical homogeneity for use in high-resolution optics, high-powered lasers, and fiber-optic transmission lines.

The TRW executive listed a wide range of low-gravity electronic products, including semiconductors, integrated circuit chips, magnetic switches, relays, magnetic detectors, ultrasonic and optical frequency filters, superconductors, high-power rectifiers, radiation detectors, light-emitting diodes; high index of refraction glass, advanced lenses and mirrors, and biological products.

Among the biological and pharmaceutical potentials of space processing are human cell purification, vaccines, enzymes (such as urokinase, which dissolves blood clots), immunoglobulin, hormones, and techniques of isolating red blood cell fractions.

In the realm of structural materials, higher-strength alloys could be produced in a low-gravity environment, along with better lubricants. Superior-strength metals for turbine blades would have averted many of the test failures in the development of the Shuttle's main engine.

In addition to levitation, the absence of gravity in orbital free fall allows better mixing and distribution in elements making up compounds, such as alloys and ceramics.

Beyond NASA

At the opening of the Shuttle era, the main thrust of civilian space programs has been in the realm of scientific investigation of the Solar System and the development of public utility satellites, such as communications, weather, and Earth monitoring machines (Landsat and Seasat). But we have yet to develop a space economy.

Except for the communications satellites, the investment of about $100 billion in space technology has had little perceptible impact on the American economy. Perhaps it may be compared with the investments of Queen Isabella in Columbus and of Charles V in Magellan. The immediate return to the economy of Spain was small, but the unforeseen potential was vast indeed.

History shows us how illusory it is to predict the future by extrapolating present trends; always the unforeseen or the unrecognized development alters the outcome. Who at the end of the fifteenth century could have predicted the rise of a new world superpower in 500 years?

formulated since the Space Task Group proposals of 1969.

Consequently, it was clear from the symposium that the United States was operating a space program on the basis of ad hoc development to meet specific, usually ephemeral, situations, such as the crash effort to land men on the Moon. The Shuttle too was an ad hoc development. At the outset it was promoted merely as a cheaper substitute for expendable launch vehicles, with other uses added on as they became feasible.

Both Senators Stevenson and Schmitt were attempting to formulate a new space policy, but neither seemed to be getting anywhere. The big stumbling block was the Office of Management and Budget, which, Stevenson said, reflected the president's priorities in considering requests from Congress. "It's a process which attaches very little value to most subjective values," the Illinois senator said. "Those that are easily quantified tend to assume the highest priorities in our government; value such as scientific research for which the return is not always predictable . . . tends to lose out in the process which is an objective process."[11]

Carl Sagan, Cornell University astronomer and astrophysicist, estimating that the colonization of Mars would cost at least $1 trillion, cited the cost advantage of automated, robotic devices in the exploration of the Solar System. Beyond those was the *Space Telescope*, which, he said, has the potential of answering the most fundamental cosmological questions—such as whether the universe will continue to expand forever or go into reverse and contract to the primeval core from which it started, presumably with a big bang.

Conferees were asked to speculate on where we would be in space ten years from the date of the symposium—from July 19, 1979. No one hesitated in projecting his views.

Hinners: We've just launched the sample return mission to Mars. We will be starting to prepare for sending the modules to establish the first human base.

George Jeffs, Rockwell International: I have high hopes that we can address some of the problems associated . . . with the production of energy from space.

Fletcher: I think we will have ordered 10 Space Shuttles by then.

Sagan: Well, the choice, as H. G. Wells has said, is the universe or nothing. I hope that in 10 years we have the wisdom to choose the universe.[12]

Countdown

By summer 1980, the first manned orbital flight had been postponed nine times. In 1972, the space agency had promised Congress that the first mission would be launched in March 1978. But by early 1977, when the glide tests with *Enterprise* began, the prospect of an orbital test flight had slipped to March 1979.

A succession of postponements followed: September 28, 1979; November 30, 1979; December 9, 1979; March 30, 1980; June–July 1980; September–October 1980; November 30, 1980; March 1981.

The slippage, the result of repeated main engine and heat shield failures, seemed to be endless until the summer of 1980, when enough progress had been made on both fronts to give NASA confidence that these problems could be solved within a year or less.

On August 1, 1980, administrator Robert A. Frosch announced that Orbiter 102, *Columbia*, would be launched in March 1981. There seemed to be enough maturation in engine development to justify it. NASA

headquarters actually set March 10 as the target date. It was sufficiently distant to be credible.

Although Shuttle delays played no part at all in the presidential election campaigns of 1980, the NASA administration exhibited a political compulsion to demonstrate that the decade-long project was approaching a conclusion. It was clear that the agency was suffering from an acute case of diminished credibility.

The list of postponements and the agency's inability to forecast its funding requirements to Congress had eroded NASA's old prestige. The "can-do" image of an agency that had put men on the Moon was fading. There was talk of giving NASA programs to the Department of Defense, although two DOD manned space projects, Dynasoar and the Manned Orbital Laboratory, had failed to reach the launch pad.

The Shuttle directorate that had snared congressional support for the program by making it appear a piece of cake now became the victim of that illusion. If it was so

easy, why couldn't NASA pull it off? Serial postponements and budget miscalculations left an impression of gross ineptitude. Early in 1980, Frosch and his deputy, Alan M. Lovelace, again begged a supplement—now $300 million—to their current budget only a few months after having assured the congressional committees that supplementary funds would not be needed. Their fiscal 1981 budget called for an increase of more than $800 million for the Shuttle.

In 1971 dollars, Shuttle development cost had risen 22 percent, but in terms of 1978 dollars, it had soared 64 percent. Instead of the $5.15-billion cost estimated in 1972, it was going to cost $8.7 billion in current money to make the first orbital test flight. That did not cover the cost of three additional orbiters in the fleet. Of course, inflation created most of the dollar overrun, but it was still a stiff initiation fee for America to rejoin the manned space club.

Defense to the Rescue

Inasmuch as the program had long since passed the point of no return, Congress had no choice but to find additional money for it. Aerospace industry lobbying helped some, but the main impetus for continuing with Shuttle development came from the Pentagon.

Secretary of Defense Harold Brown told the Senate Committee on Commerce, Science, and Transportation during hearings in February 1980 that the Shuttle was a critical element in American defense planning. It was especially important, he said, in enabling the United States to monitor a Strategic Arms Limitation Treaty (SALT) by means of surveillance satellites.

The Shuttle, now perceived as an important element of national defense, was acquiring the immunity to retrenchment of an essential weapons system. Its potential of tapping the resources of the Solar System was made to seem secondary in the light of its emerging military significance. In World War III, space might be the high ground.

The military potential of the Shuttle ranged far beyond its role as a satellite ferry. It was the first manned space vehicle equipped to seize a satellite within its altitude range of 600 miles. Its capability for reconnaissance was unmatched. It could be armed with missiles and laser cannon against killer satellites.

The Pentagon people were convinced that the Soviet Union had developed killer satellites. The Shuttle might be the very thing to combat them. For their part, the Soviets agreed that the Shuttle had definite military potential and went so far to suggest that this was the main reason the United States was building it. Soviet propaganda characterized the Shuttle as a destabilizing force. NASA ridiculed the idea, insisting its mission was peaceful, scientific, commercial. But Defense plans for using it tended to blur that view.

A Dismal Outlook

Despite Frosch's insistence in the summer of 1980 that major technical problems delaying the Shuttle had been resolved, the pattern of failure persisted in engine testing.

At the end of 1979, the Shuttle main engines were being run for full-duration tests. On this 550-second run December 17, 1979, the main propulsion test article reached all test goals. Flocks of birds roosting around the test stand at the National Space Technology Laboratories, Bay St. Louis, Mississippi, were scattered by the roar of the engines and clouds of steam.

(NASA)

Following a successful test of a three-engine cluster in December 1979, an identical test failed on February 1, 1980. The engines cut off 46 seconds into a 550-second run when the high-pressure oxygen turbopump overheated in engine 3. Then successful cluster tests were run late in February and in March, but on April 16, fuel pump overheating in engine 2 cut off a three-engine test after 5 seconds.

After tracing a series of engine failures in 1979 to improper welding wire, Rockwell and NASA engineers discovered that the same wire had been used in the three engines installed in *Columbia:* 2005, 2006, and 2007. They were taken out of the orbiter and shipped back to Rocketdyne in California for rewelding, along with modifications of the turbopumps, valves, and nozzles to correct difficulties that had shown up on single-engine tests.

Columbia's engines were returned to the National Space Technology Laboratory at Bay St. Louis for another round of acceptance tests after the modifications were made. Each engine was fired individually during

In January 1980, while *Columbia* was in the Orbiter Processing Facility, its engines were removed for welding repairs to the hydrogen manifolds of their nozzle cooling systems.
(NASA)

While engine tests were running at Bay St. Louis, *Columbia* was being prepared for its first mission in the Orbiter Processing Facility at Kennedy Space Center. This view of the orbiter's payload bay shows the port side door open. Because the hinges were not designed to bear the weight of the door while the orbiter is horizontal on the ground, a counterweight supports the door, simulating the microgravity effect in orbital flight.   (NASA)

June 1980. The three were then shipped to Florida to be reinstalled in the spaceship.

Meanwhile, the ninth successful firing of the three-engine main propulsion system was completed at Bay St. Louis during June. There were three more cluster firings to go in the Main Propulsion Test Article (MPTA) series before the orbiter's main propulsion system could be certified for launch.*

Columbia's engines had not been fired as a cluster, although this had been recommended by the Ad Hoc Committee of engineers that reviewed engine development for the Senate space committee. NASA argued that cluster firings of engines at the same stage of development as those in *Columbia* were sufficient guarantee the system would work as advertised. However, the agency inserted into prelaunch preparations a 20-second firing of *Columbia*'s engines while the shuttle was on the pad.

The engine test program was looking good until July, when the tenth cluster firing failed. A hole burned through the preburner of one engine and all three shut down in seconds. Later in the month, fire broke out in the oxygen pump of engine 0010, a backup engine for *Columbia,* as it was undergoing preliminary acceptance testing. The engine was sent back to the factory to correct faulty fuel flow. Another three-engine test firing failed in November, when the high-pressure fuel pump overheated in engine 2. A hole had been burned in the nozzle when a brazed joint failed and disrupted coolant flow.

All these failures were replays of familiar problems. By the end of 1980, the contractor knew how to fix them. The eleventh cluster firing ran 591 seconds on December 4, 1980, and the twelfth ran 629 seconds on January 17, 1981, completing the engine test program. The final test ran more than 119 seconds longer than the 8 minutes, 30 seconds required to boost *Columbia* into orbit.

With this test, cumulative engine firing time of both single and clustered engines reached 100,000 seconds—20,000 more than the goal of 80,000, which Shuttle directors had set in 1979 as a condition for launching the first manned orbital flight.

Out of the Hangar

After 20 months of trial and error, *Columbia*'s heat shield was nearing completion. The NASA Tile Certification Committee at headquarters advised that 4,500 heat-critical tiles be strengthened by a process that would increase their density. Most of these were the six-inch-square black tiles on the underside of the orbiter, where heat loads were expected to be high.

The committee recommended that a tile repair kit and a manned maneuvering unit (MMU) be carried on the first flight. If tiles were ripped off during launch, crew members could make emergency repairs by climbing outside the vessel in space suits, wearing their MMUs. They could fill the holes with an epoxy resin that would protect the aluminum skin through reentry.

The MMU converted a space-suited astronaut into a virtual spacecraft, but its Buck Rogers–Batman–Super-

A Massachusetts Institute of Technology student, Dave Akin, tests a Manned Manuevering Unit (MMU) built by students at MIT in the Neutral Buoyancy Simulator, a 1.3-million-gallon water tank, at the Marshall Space Flight Center. Worn on the back, the MMU allows the space-suited astronaut to fly independently in space without being tethered to the Shuttle.
(NASA)

man appeal held no charm for Young and Crippen. They expected to be busy simply learning how the spaceship worked. Frosch finally vetoed the idea as too risky. Besides, it would delay the March 1981 launch target date by weeks or months.

In November, the engineers in charge of *Columbia*'s processing—Kenneth Kleinknecht, Robert Gray, and KSC Director Richard Smith—announced that the heat shield was installed. Only narrow gaps between some tiles remained to be filled, and that could be done in the Vehicle Assembly Building (VAB) or even on the launch pad.

While *Columbia* was in the Orbiter Processing Facility, the prime crew paid it a visit to rehearse the mission from the flight deck. Here John W. Young, commander, and Robert L. Crippen, pilot, arrive at the OPF January 10, 1980, as a smiling guard waits for them to hand over their badges for admittance.   (NASA)

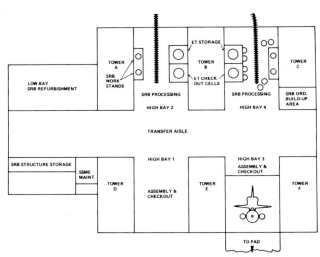

Shuttle facilities layout in the Vehicle Assembly Building.
(NASA)

High Bay 2 being altered from its Saturn-Apollo configuration to accommodate solid rocket boosters and external tank processing.   (NASA)

On November 24, 1980, *Columbia* was rolled out of its hangar—the Orbiter Processing Facility (OPF)—and into the VAB to be mated with the external tank and solid rocket boosters.

The rollout was a major event in aerospace circles. Initially scheduled for noon, November 23, it was held up for 18 hours by minor problems. At long last, *Columbia* emerged tail first from the hangar at 6 P.M. eastern standard time. It had been there since its arrival at the space center March 24, 1979.

The sun was setting beyond the Indian River as the 122-foot delta-winged spaceship was pushed onto the concrete apron in front of the OPF by a small tractor. In the fading light, the ship shone in its black and white tile dress. It exhibited an ungainly beauty, moving silently on its rubber-tired landing wheels as the tractor swung it around to face the Atlantic Ocean.

Dusk deepened and floodlights went on. An audi-

ence of about 1,000 space workers and their families crowded up to a barricade of yellow rope to watch. Probably about half of them were tile installers, temporary workers whose job had ended. Many had received their pink slips a day or two earlier.

A young woman in a Rockwell coverall, her long hair flying in the evening breeze, waved her dismissal notice

A Shuttle external tank, 154.2 feet long, is trundled toward the Vehicle Assembly Building at Kennedy Space Center, where it is to be mated with twin solid rocket boosters and the orbiter.   (Martin Marietta)

and screamed, "Hail, *Columbia!* We who are unemployed salute you!"

There was a ripple of laughter, which swelled to a wild cheer and applause.

The spaceship remained parked for a few minutes while a rotary sweeper cleared a path on the debris-strewn concrete apron to the towering VAB. Lights gleamed in the cavernous interior of the assembly building.

"Let's go, let's go," someone yelled.

Director Smith waved a hand and pointed to the sweeper. "That's all we need now," he bellowed into the public address. "A nail in the tire."

The orbiter began moving again toward the VAB, just a football field away. The crowd followed. *Columbia* was towed into the transfer aisle through the north door of the cathedral structure. (Before it was fully air conditioned, clouds used to form at the 500-foot level and drop rain on the floor.)

The aft assemblies of *Columbia*'s solid rocket boosters are hoisted onto the mobile launcher platform in the assembly bay.   (NASA)

Segments of a solid rocket booster are stacked in a high bay.
(NASA)

After the erection of the twin solid rocket boosters and the external tank on the mobile launcher platform, *Columbia* was moved from the Orbiter Processing Facility to the Vehicle Assembly Building to be mated with them November 25, 1980. *Columbia* is lifted (left) from the floor of the VAB by overhead hoist. After it is swung over a large structural beam and lowered to a position just above the platform, *Columbia*, shown here (right) still in its support sling, is mated with the boosters and the external tank in High Bay 3 of the VAB.   (NASA)

Before midnight, lifting beams and erection slings were installed around the vessel. It was lifted nose up, landing gear retracted, to a vertical position by a pair of huge cranes. Then it was moved to the assembly area in High Bay 1 and bolted to the External Tank, to which the Solid Rocket Boosters had been attached previously.

Fully assembled, Space Shuttle No. 1 rested on the aft skirts of the big boosters atop a red-painted mobile launcher platform. Within ten days, all of the electrical connections had been made. The Shuttle was ready for its first ground test as an integrated flight system.

The test began December 4, 1980, and continued for two weeks. The early part was routine checkout of connections. The flight-control phase began December 10.

It was supposed to verify the ability of the flight computers to control the ascent path by gimballing the SRB rocket nozzles during the first two minutes of flight.

Later phases of the ascent to orbit were run the following week. Four simulated missions in which the prime crew, Young and Crippen, and the backup crew, Joe Engle and Richard H. Truly, repeated ascent and descent were carried out.

After the first simulation of the descent from orbit, John Young, looking tired and serious, explained what it would take to fly the orbiter. Human reaction time, he said, is not fast enough to control it during the high-speed phases of the descent. It takes the computers to do that.

Earlier spacecraft used flight computers—''black

boxes"—that could handle some piloting functions better than a human pilot. They could react faster. In Mercury, Gemini, and Apollo, the pilot could bypass the black boxes and apply manual control to fly the vehicle.

That was not the case on the orbiter. The computers, blandly referred to as the "data processing system," fly the vessel, and the pilot can control it only through them. As mentioned in the first chapter, there is enough redundancy in the data processing system's five flight computers to make a complete system failure unlikely in the extreme. But if that should happen, the pilot would be helpless. There would be no way that he could control the orbiter, and it would crash.

Swan Song

Early in October 1980, Robert A. Frosch announced his resignation as NASA Administrator. He did so in order to become the first president of the newly organized American Association of Engineering Societies. It was an opportunity, he said, that he simply could not pass up.

Frosch's resignation took effect January 20, 1981 with the change in the national administration, and his deputy, Alan M. Lovelace, took over as interim administrator. President Reagan then nominated James M. Beggs, 57, for the top NASA post. Beggs, a veteran aerospace executive, had served in NASA during the Nixon administration. He was confirmed and took office in June.

What space policy the new administration would support was not at all clear. President Reagan had not given any clue to his attitude on space exploration and development during the political campaign of 1980. For that matter, neither had Mr. Carter.

As far as the rhetoric of the campaign went, the future of America in space was simply ignored by both candidates. It appeared reasonable that the new president would see to it that the Shuttle became operational, despite the administration's drive to reduce the general budget, because of the vehicle's military potential.

Except for the Shuttle, the space program of the United States was at a standstill. The Carter administration had contributed little to it beyond approving continuation of the soundings of the outer planets and an atmosphere probe and orbital radar scan of Venus, which depicted its surface features on a broad scale. Mars had been abandoned after the magnificent exploits of the Viking program. Nothing was planned for the Moon.

Frosch visited the Kennedy Space Center early in December and held a press conference that provided a platform for an impromptu farewell address. He recommended that NASA use the Shuttle to learn how to build large structures in orbit, such as a space station.

He recalled that the original purpose of the Shuttle had been to provide commuter service to an orbiting space station. "I could characterize the last several years as a set of attempts to break out of what has been a constraint situation," he said. "We didn't succeed in finding some overriding, 'let's go to the Moon' kind of goal. That's partly because it is not a 'let's go to the Moon' sort of era."

Still, it was an era in which "we got a lot of things going." We got science going, and we're trying to build a brand new space transportation system."

In addition to the brass ring of a better-paying job as it came around, Frosch admitted that the jeopardy of remaining in an executive agency during an election year had been a factor in his decision to leave, even though he was doing so at the climax of the development effort. "As it turned out, I presumably would have had to leave in any case."

Like other top jobs in Washington, the NASA administrator post is a political appointment. The president usually selects a member of his own party. President Eisenhower picked T. Keith Glennan, a Republican. John F. Kennedy appointed James E. Webb, a Democrat and the only nontechnical administrator. Webb's role, however, was to sell the Moon to the American taxpayer and to Congress. And he did.

Webb retired and was succeeded by another Democrat, Thomas O. Paine, who served through the Johnson administration. Richard Nixon replaced him with a Republican, James C. Fletcher.

The political allegiance of these executives probably was not of first importance. Each president wanted a team player, an administrator who would manage the space program in consonance with the president's policy.

In the context of the missile gap and the space race, Kennedy needed a super salesman. Webb fulfilled that role admirably. Earlier, Glennan, who had been president of Western Reserve University, reflected Eisenhower's cautious development policy in 1958–1959 and particularly the wartime leader's antipathy toward military control of manned spaceflight and space research.

Paine, who succeeded Webb in the gung-ho period of lunar exploration, sought to continue the bold approach by implementing the 1969 Space Task Group's manned spaceflight recommendations. One of them called for a manned landing on Mars, circa 1985. But that goal was shot down by the Vietnam war and its effect in tarring manned spaceflight with military adventurism. Paine who had been a vice president of the General Electric Company, found himself politically isolated in the administration of Richard Nixon. His resignation and return to his old job marked the end of manned interplanetary flight ambitions for this century.

Fletcher, who succeeded Paine in the Nixon period of retreat from the Moon, had been president of the University of Utah. He presided over the early years of Shuttle development and faithfully carried out the money constraint policy that was to create delay later on.

So did Frosch. A former Assistant Secretary of the Navy, he followed orders, too. In his swan song to the press, he expressed regret that he had not tried more vigorously to break out of the OMB's funding constraints. He should have rocked the boat, he said.

Interlude at Whiskey Creek

Although it lacked the glamor of the lunar landings, the building of the Shuttle was a benchmark in engineering science. Its electronic brain and avionics, the heat shield and the main engines, were dramatic advances. Unhappily for the NASA directorate, they were perceived by the communications media, Congress, and the public only in terms of their difficulties and cost overruns.

History may some day conclude that the Shuttle was as new to the world as Noah's Ark. Nothing like it ever had been built before. But the fact seemed to impress no one. Alan Shepard's little Mercury capsule, primitive by comparison, evoked the awe of a miracle. But the long-delayed Space Shuttle Transportation System was a source of embarrassment.

The propulsion component of the system that evolved smoothly was the solid rocket boosters. They were built upon a decade of established technology. Only one aspect of their operation was new and untried: their recovery from the sea and subsequent reuse. As mentioned earlier, the twin SRBs were designed to be recovered from the sea, returned to the launch site, refurbished, and reloaded for another go. Shuttle directors estimated that the rocket casings could be reused 20 times.

How much money reuse would save in view of the cost of recovery and refurbishment was speculative. But reusing the SRBs lent credibility to the reusable character of the vehicle. Only the big propellant tank was expendable.

The recovery process was complex. At burnout, two minutes after liftoff, the SRBs would separate from the orbiter at 27.5 miles altitude by means of small rockets thrusting them away. They would continue to coast upward for about 70 seconds to 41.6 miles and then arc

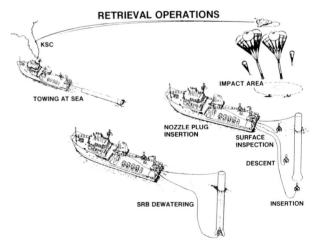

RETRIEVAL OPERATIONS

KSC

TOWING AT SEA

IMPACT AREA

NOZZLE PLUG INSERTION

SURFACE INSPECTION

DESCENT

SRB DEWATERING

INSERTION

Two minutes and five seconds after launch, the solid rocket boosters burn out and are dropped into the ocean. Their retrieval for reuse is a complex operation performed by specially built ships. This diagram summarizes the retrieval process.   (United Space Boosters)

over to begin a long free fall to an ocean impact area about 160 miles downrange from the launch site, Pad 39-A.

As *Columbia* and its big tank continued toward orbital altitude, the SRBs' free fall would continue until the burned-out rockets reached 15,400 feet. An altitude switch would then free the nose cap of each rocket, allowing a pilot parachute to come out of the casing. The pilot chute would bring forth a 54-foot-diameter drogue parachute, which was designed to stabilize the falling rocket. At 6,600 feet, a section of the rocket called the frustrum would separate, allowing the three main parachutes, each 115 feet in diameter, to emerge and unfurl.

It was calculated that each rocket would hit the water

One of the retrieval ships is United Technology Corp.'s *Liberty*, specially designed for retrieving the booster casings and parachutes from the sea. The retrieval is conducted by UTC's subsidiary, United Space Boosters, Inc.

(United Space Boosters)

about six seconds short of seven minutes after liftoff, with an impact of 59.3 miles per hour. There would be a big splash but no damage.

Standing by the predetermined splash zone would be two retrieval ships named *Freedom* and *Liberty*. They were designed and built for the job by the recovery-refurbishment contractor, United Space Boosters, Inc., a subsidiary of United Technologies Corporation. Each of these specialized vessels was 150 feet in length and was equipped with cranes and deck reels to lift the frustrum and parachutes out of the water and to tow the rocket casing back to Port Canaveral and the space center. The parachutes were to be reeled in like fishing nets and the heavy metal frustrum lifted over the side of the recovery ship by a deck crane.

But the complicated part of the operation would be recovery of the rocket case. It weighed almost 88 tons and was nearly as long as the recovery ship. It would be one-third to one-half filled with sea water when encountered and would be afloat in a nearly vertical position, nozzle down, thanks to air trapped in the upper tanks.

In order to tow the rockets back to shore, the recovery force must pump the sea water out of the casing so that it would fall over to a horizontal floating position. Next, the nozzle must be corked so that no additional seawa-

A diver (foreground) inspects the aft skirt of a solid rocket booster mock-up during a retrieval test at Port Canaveral.

(NASA)

ter would pour in. The nozzle would take a big cork—the diameter ranged from 54 inches at the throat to 148 inches at the bell. The system for plugging it was demonstrated in 1978 by a $1.5 million invention called the Solid Rocket Booster Dewatering Set (SRBDS), or nozzle plug.

The plug was developed in five years by the Naval Ocean Systems Center, an agency with a history of recovering things in the sea from nuclear bombs to submarines. A Canadian firm, International Hydrodynamics, Ltd., was awarded a NASA contract to build two plugs at a cost of $1,022,315 each.

I watched a demonstration of the plug during one of its early tests in the winter of 1978. The test site was the Whiskey Creek area of Port Everglades at Ft. Lauderdale. It was not a prepossessing locale. The oily waters of the creek were strewn with beer cans, pop bottles, broken vegetable crates and pieces of furniture. The demonstration had been scheduled at sea, but various malfunctions delayed it until late afternoon. A rough sea came up and confined the ship to the rubbish-littered waters around the dock.

The plug was a torpedo-shaped robot, 14.5 feet tall,

The radio-guided nozzle plug developed by the U.S. Navy for booster casing recovery.   (NASA)

generally cylindrical, with arms that extended to 7.5 feet in diameter to lock the tip into the SRB's nozzle throat. It weighed about 3,500 pounds and had to be hoisted over the side of the ship and lowered into the water by a derrick.

All this was done about mid-afternoon of a windy, cloudy day. In the water, the black-and-white goblin bobbed up and down amid the refuse like an expectant seal. Attached to it was a 600-foot umbilical cord that carried electric power from the ship to run the plug's six propeller-type thrusters, two brillant searchlights, and four closed-circuit television cameras.

Suddenly, the plug began moving rapidly through the water in its vertical position. It looked like a creature out of a science fiction movie. Its television eyes gleamed in its turret as the sun broke through briefly and illuminated the scene with a wan yellow light.

The robot was being operated from a shipboard console in a closet just aft of the cookhouse, from which emanated the waft of frying pork chops. Dinner time was near, and soon it would be dark.

As I squeezed into the cubby to watch, the engineer was testing control of forward and sideways motion with a hand controller, all the while monitoring the plug's movements in a small TV screen. He could direct the plug through three dimensions with ease, and its response was remarkably quick, smooth, and precise.

Two years later, in December 1980, the nozzle plug was tested in the Atlantic Ocean off Cape Canaveral. A mock-up of the SRB was its target. Hoisted over the side of the ship, the plug swam in a vertical position to the booster, which floated nozzle down in the water. Reaching the mock-up, the plug dived by simply dropping straight down. It could reach a depth of 135 feet. Its powerful searchlights went on and showed the underwater segment of the rocket mock-up and the nozzle.

Watching the nozzle throat on the television screen, the shipboard controller steered the nose of the plug into the nozzle aperture. A light went on in the console, indicating that the plug was "docked." At a signal from the console, the plug's three metal arms were extruded to their full diameter, locking the plug into place inside the nozzle.

The controller turned on an air pump that forced air through a hose in the umbilical line at a pressure of 75 pounds per square inch. Seawater poured out of the rocket and it settled slowly toward a horizontal float.

Another signal from the console caused a rubber bag to inflate around the plug and seal the throat. Some residual water in the rocket case was pumped out through a short hose. Now horizontal, the rocket was ready for the towline.

Under actual conditions, each ship would retrieve one SRB and its parachutes. The parachutes would be washed and dried at the launch site in a huge laundry. The casings would be scrubbed and shipped to Thiokol in Utah to be reloaded.

Recovering the boosters at sea would be a new experience. Would it work? Although not essential to the flight, the success of the recovery on the first test flights would indicate whether the Shuttle would ultimately become cost effective.

Rollout

On December 29, 1980, *Columbia* was rolled out of the Vehicle Assembly Building fully mated with its propellant tank and solid rocket boosters. The three components of the Shuttle weighed approximately 6 million pounds and were mounted on a mobile launch platform. The platform in turn was carried slowly out of the building by a giant tractor-crawler, a machine designed in the 1960s to carry the Saturn 5–Apollo to the launch pad.

It took nearly six hours for the crawler to negotiate the three miles between the VAB and Launch Complex 39. Starting at noon, the journey was completed by dusk, when the shuttle on its launch pedestal was secured to the pad and a steel access arm was extended to enable the crew to enter the orbiter's cabin hatch.

The next day, the service structure on the pad, a vertical tower of steel girders, was moved into place on railroad tracks so that it enclosed *Columbia* like a steel web. A series of prelaunch tests began.

Flight simulations, electrical connections, and pad tests were made. On January 22, 1981, the crucial propellant

loading test was started. In the Saturn program 15 years earlier, propellant loading had been a major problem, causing weeks of delay. No one was sure what might occur now.

Approximately 139,620 gallons of liquid oxygen and 378,380 gallons of liquid hydrogen were pumped into the 154-foot-tall propellant tank. These amounts approximated the propellant load for the first manned orbital flight, planned in March.

The area around the launch pad was cleared for a radius of three miles, and a popular beach area adjacent to the space center was closed to tourists. As the tank

Columbia emerges from the Vehicle Assembly Building December 29, 1980, aboard the tractor transporter, or "crawler," which hauled the assembled Shuttle on its mobile launch platform 3.5 miles to Launch Complex 39-A. In the middle distance is the launch control center, the "brain" of the launching operation. The four-story structure has two firing rooms equipped for Shuttle launches and two firing rooms in reserve.   (NASA)

Moving eastward along the crawler way toward the launch pad, *Columbia*'s progress is so slow (about one-half mile an hour) that a busload of photographers has ample time to cross its path. The 6-million-pound crawler, developed to carry the Saturn 5 moon rocket and the Apollo spacecraft to the pad, follows a roadway 130 feet wide. (NASA)

Columbia reaches the launch pad on the afternoon of December 29, 1980, for its initial flight, designated STS-1. As darkness falls on the short winter day, the launch complex becomes ablaze with floodlights. (NASA)

was filled with propellant, it became a potential bomb with an explosive force of 3.15 kilotons.*

The possibility of such an accident had to be considered, because the launch of *Columbia* would be the first attempt to ignite engines fueled by hydrogen on the ground—the first attempt anywhere in the world. A blowup at launch would probably cause the SRBs to explode also and spread toxic fumes over the launch site, with the effect of poison gas. The fumes would reach the press site, 3.5 miles away. During Saturn 5 launches, which were considerably more benign in terms of po-

tential hazard than a Shuttle launch, observers and the press were kept at a distance of 3.5 miles from the pad. The launch team was protected by blast-proof windows and shutters in the firing room. The pad crew was evacuated late in the countdown, leaving only the flight crew. These procedures were continued for Shuttle launches.

The aircraft-style ejection seats provided the commander and pilot on the first four test flights could not be used while the orbiter was on the pad. A sudden pad explosion would thus be fatal to the flight crew. With advance warning, however, the astronauts could get out of the orbiter before the access arm was swung away and descend to ground shelters via the slide wire escape route.

*See ch. 3. This was the "worst case" estimate of NASA's 1977 Environmental Impact Statement.

The Cryogenic Crisis

The fuel and oxidizer loading test appeared to be successful until inspection showed that 32 panels of cork-epoxy insulation had peeled off the portion of the propellant tank where hydrogen is stored. The denuded area was about 8 square feet.

Tank insulation was designed for dual duty. In addition to warding off frictional heat during the ascent, it prevents the super-cold propellant from freezing moisture on the outer tank wall. The hydrogen is stored inside at $-423°$ F. and oxygen at $-297°$ F. Ice forming on the outer tank wall while the Shuttle is on the pad is a serious potential hazard. Chunks that break off during launch could damage the brittle heat shield tiles on the forward fuselage.

The cause of the adhesive failure became a target of intense investigation at the space center and the New Orleans (Michoud) factory where Martin Marietta engineers assembled the tanks. It also became a new and unexpected "pacing item" for the launch. Until the cause could be found and corrected, *Columbia* was stalled. The March launch date withered away, just as all the others had.

NASA was not alone with launch delays. The ambitious effort of the European Space Agency (ESA) to put an old-style expendable rocket called Ariane into operation as a competitor to the Shuttle became stalled also. Ariane was experiencing engine trouble. Based on conventional launch vehicle technology, as I have related earlier, Ariane's development seemed to be leading that of the Shuttle until Murphy's Law intervened.

Following a successful first launch from France's Guiana Space Center at Kourou on December 24, 1979, Ariane failed on the second launch attempt, on May 24, 1980, when the first stage engine system malfunctioned. A third launch attempt was delayed until mid-1981. The problem was attributed to high-frequency vibration (pogo), which was more severe than expected. American rocket engine developers had been fighting pogo for 20 years and had managed to subdue it in the Space Shuttle Main Engine.

These up-and-down, pumpinglike oscillations were the plague of liquid-propellant rocket engines. Pogo nearly wrecked the early test flights of the Saturn 5 in 1967–1968. Despite an intensive effort to suppress these oscillations in the SSME, NASA and Rocketdyne engineers could not be sure the suppressors they had designed would work until an actual launch. As H. E. Whitacre of the Johnson Space Center told a news conference in November 1980; "This [pogo] is such a dynamic problem and it is so difficult to analyze that the only real way that you ever find out that you've got a good, safe margin on pogo is you go out and fly it."

And that was the formula for determining a good, safe margin for every other Shuttle problem, from the engines to the heat shield.

The Flight Plan

The plan of the first manned orbital flight called for launching *Columbia* at an inclination of 40.3°. Its flight path thus would follow a ground track between 40.3° N. and 40.3° S. latitude. On several of its 36 revolutions during the 54-hour mission, it would pass over Philadelphia, Madrid, Ankara, Peking, and Melbourne, Australia. At 150 nautical miles altitude, it would be visible as a bright, moving star at dusk and dawn.

Two minutes after liftoff, the SRBs would cut off and be ejected. The SSME system would continue blasting *Columbia* toward orbit for an additional 6 minutes, 32 seconds. When the main engines cut off, the big propellant tank would be dropped. At this point, *Columbia* would be entering a low, oval orbit of 13 by 80 nautical miles. Two minutes later, the twin orbital maneuvering engines would ignite and fire for 1 minute, 50 seconds, lifting the flight path to 150 by 57 nautical miles. After an interval of 33 minutes, 30 seconds, the space engines would fire again, to circularize the orbit at 150 nautical miles, or 172.5 statute miles.

On a launch due east from Florida, the main engines could put *Columbia* into a 115-mile orbit without assistance of the orbital maneuvering engines, provided cargo was light. However, ascent procedure called for throttling back the engines, cutting them off, and dropping the external tank before the ship reached orbital speed. This procedure would prevent the tank from going into orbit and becoming a hazard to navigation. The tank was dropped at a point on the ascent that would allow it to fall into the Indian Ocean.

Descent was scheduled to start at 53 hours, 23 minutes after liftoff over the Indian Ocean between Africa and India. *Columbia* would then commence its long glide across the Pacific Ocean to California. It would enter the atmosphere at 400,000 feet altitude to the west of Wake Island at 53 hours, 59 minutes, and 32 minutes later, it would land on Rogers Dry Lake.

Abort

The possibility that one of the orbiter's main engines will fail during the launch required the Shuttle directorate to devise several modes of aborting the mission. Each is designed to save the crew and preserve the vehicle.

If an engine fails in the first 4 minutes, 23 seconds of the launch, the crew would try to fly back to the launch site on the remaining two. The SRBs would have been jettisoned at 2 minutes. At a predetermined point on the return flight, the remaining two engines would be shut down and the external tank would be dropped into the Atlantic Ocean. The crew would then fly the orbiter as a glider back to Florida and land at the Kennedy Space Center. The return-to-launch-site abort was estimated to take 22 minutes.

If engine failure occurred after 4 minutes, 23 seconds, several abort options would be available. By then the orbiter would have reached the point of no return. With two engines out, it could cross the Atlantic on the remaining engine and land at the U.S. Naval Air Station at Rota, Spain, or at the Dakar airport in Senegal.* If a single engine failure occurred at the point of no return, the remaning two engines would enable the ship to make one circuit of the Earth with the assistance of the orbital maneuvering engines and land at Edwards or at White

*On a trans-Atlantic abort, the landing site would be determined by the inclination of the ascent path to the equator.

Sands, New Mexico. The crew would execute the deorbit maneuver over the Indian Ocean as on a nominal flight and start down. However, the once-around option might put the orbiter on a descent path too far south to enable it to land on Rogers Dry Lake. It would then have to put down at an emergency landing strip at White Sands.

A second option is available if single engine failure occurs more than 20 seconds after the point of no return. The crew can then reach a 105-nautical-mile circular orbit on two engines and fly the full mission at that altitude, with a landing in California.

If more than one main engine fails during launch, the crew ejects from the cockpit at low altitude and the orbiter goes into the sea.

The Postponement Postponed

On February 16, 1981, the launch crew started a 58-hour "wet" countdown, which was to culminate in a 20-second test firing of the main engines while *Columbia* was securely bolted to the pad.

Again the external tank was filled with propellant. Since the orbiter was not going anywhere, the gap in the tank's insulation could be tolerated. By the morning of February 20, the countdown was completed and the engines came to life with a flash, an earth-shaking roar, and a cumulus cloud of steam. A million pounds of thrust shook the orbiter and its solid rocket boosters.

Suddenly, the engines cut off, the roar dwindled into

Just before Shuttle flights, the mission commander and pilot practice approaches and landings in a Shuttle trainer aircraft at Kennedy Space Center. The trainer shown here is a modified Gulfstream II, a twin-engine executive jet aircraft, built for NASA by the Grumman Aerospace Corp.   (Grumman)

diminishing echoes, and the cloud rolled off toward the ocean. *Columbia* had passed its final test prior to launch.

The flight crew, Young and Crippen, observed it from a safe altitude of 3,000 feet as they flew over the launch pad in a Grumman Gulfstream jet they had been using to practice dead-stick landings. Everything looked fine to them. The next time those engines fired, they would be sitting right on top of the fire.

But the "next time" remained uncertain. NASA had reset the launch to the week of April 5, but the task of restoring the insulation on the external tank was delayed by a sudden strike of 1,000 employees of Boeing Services International, a major contractor at the space center. Executive personnel and strikebreakers were rushed in to maintain the schedule, but the union—the International Association of Machinists—stood firm. Mediators flew in from Washington and talks started and stopped.

On March 4, I received a call from Kennedy Space Center public information advising that an announcement of the postponement of the launch would be postponed until further notice. It was the first postponement of a postponement of the season.

Liftoff

April 14, 1981. It was cool before dawn in the high desert. Three and one-half years had passed since *Enterprise* had landed here on the last of its low-altitude glide tests. Now the crowds were gathering again, this time to see *Columbia* return from space.

A caravan of headlights turned off the road and approached the tan bed of Rogers Dry Lake in a swirl of bright dust. Lights sparkled for miles along the road net of Edwards Air Force Base. Beyond it, thousands of vehicles jammed the highways.

Columbia was beginning the 34th revolution of its first orbital test flight. In 3 hours and 20 minutes, it was due to land on runway 23, which extends seven miles on the dry lake bed.

John Young and Bob Crippen had been awakened by a call from Mission Control, Houston, and a blast of country-western music. They had breakfasted on dried apricots, "mush flakes," and processed sausage and eggs. They were in exceedingly good humor this morning. *Co-*

lumbia's performance as a spaceship had exceeded their most optimistic expectations. What remained to be demonstrated was its performance as a trans-sonic glider when it reentered the atmosphere.

The voices of the crew and of the capsule communicator at Houston could be heard through public-address loudspeakers NASA had erected on steel towers in a viewing enclave beside the lake bed. Circus-style tents had been set up for state officials, some of the beautiful people from Hollywood, and Air Force brass. VIPs were arriving in busloads. A thousand television, movie and still cameras were lined up on tripods, car roofs, and trucks behind a yellow security rope.

The dialogue between *Columbia* and Houston was too terse and technical to be comprehensible to the crowds, but it provided a link between them and the distant spaceship, flying 149 nautical miles above the Earth.

Dawn came in a glow of pearly light that glimmered over the lake bed to create the mirage of water and waves.

The air warmed, the sky brightened, and the first beams of sunlight fell on a half million people standing in the desert awaiting *Columbia*.

Space Shuttle No. 1 was due to touch down at 10:22 A.M. pacific standard time, 2 days, 6 hours, and 22 minutes after it was launched from the Kennedy Space Center in Florida. It would be a moment in history, the beginning of Part Two of the space age.

Rarely in the development of civilization has an "age" had an identifiable beginning. The old stone age, the new stone age, the bronze age, the age of iron; their origins are blurred in time. Only the start of the atomic age and the space age can be dated with some precision. The advent of a reusable space transportation system was today.

The First Voyage

Delays in the launch of the Shuttle had seemed interminable, but they seemed to be coming to an end after Christmas 1980. The nation was on the verge on flying its new Space Shuttle Transportation System. *Columbia*, which would undertake the first mission (STS-1), was on the launch pad at New Year pointing at the future.

At the beginning of 1981, launch preparations acquired a momentum of their own. Not even the strike of Boeing service workers or the deaths of two Rockwell technicians slowed them perceptibly. The technicians perished of suffocation when they entered an aft fuselage chamber filled with nitrogen after a simulated launch countdown was completed March 19.*

However, launch processing procedures were taking longer than the schedule had projected. Repair and rebonding of the thermal insulation on the external propellant tank had dislocated the schedule, too. The March

---

*The accident took the lives of John Bjornstad and Forrest Cole, both 50. They entered the orbiter's aft chamber, which had been flushed with nitrogen after the countdown demonstration. The all clear had been sounded on the launch pad on the mistaken assuption that the nitrogen had been expelled from the chamber and replaced with air. Without breathing gear, Bjornstad and Cole were quickly overcome, along with four others in their inspection crew. They subsequently died, despite rescue efforts. The other four recovered.

launch date slipped into April—first April 5, then April 7, then April 10, largely to verify the fix of the insulation.

At last, a liftoff schedule appeared: 6:50 A.M. eastern standard time Friday, April 10. By then, the sun would be high enough for good launch photography and *Columbia* could make its approach and landing in California in midmorning, 54 hours, 30 minutes later.

Overnight, thousands of motorists jammed the highways leading to the gates of the Kennedy Space Center, Viewing areas had been designated for 50,000 spectators who had received car passes. Another 600,000 assembled along miles of beaches, causeways, and the banks of the Indian and Banana Rivers around the space center. The launch of a vehicle the size of the Shuttle would be easily visible in clear skies for 40 miles.

About 60 miles east of Daytona Beach, where the two NASA-leased ships, *Liberty* and *Freedom*, waited to recover the solid rocket boosters (SRBs) from the Atlantic, a Soviet trawler stood by as though to observe the recovery operation, and a U.S. Coast Guard vessel patrolled the area as though to keep an eye on the trawler.

At sunrise, miles-long lines of vehicles inched their way into the space center. The press site grandstand, built in the Saturn 5–Apollo days, was jammed with media people, some of whom actually were covering the event, others just there. Artists set up their easels and photographers their cameras on a grassy outfield, alive with sand fleas whose venom could raise itchy welts on the arms and legs and stomachs of victims allergic to it.

Three and one-half miles away, *Columbia* was poised under a plume of oxygen vapor rising from the huge propellant tank. Young and Crippen waited alertly in their ejection seats, monitoring the count.

At T (for launch time) minus 9 minutes, the count was held while the launch team at Kennedy and controllers at Houston investigated a computer discrepancy. It had appeared at T − 20 minutes, when the flight computers were switched from the prelaunch to the launch programs. At that point, communication was to be established between the redundant set of four flight computers and the fifth—the backup—computer. But it failed to happen.

The count was recycled to repeat the T − 20 minutes

event, but the backup computer failed to signal the other four. Mission rules required that all five computers be in synchronization and operating normally before liftoff.

At Houston, computer experts delved into the problem. They discovered a 40-millisecond "slew," or discrepancy, in timing between the redundant set of four computers and the backup. When the program shifted from prelaunch to launch, two of the four computers signalled the backup 40 thousandths of a second before it was time to receive the signal and respond. It recognized a discrepancy and did not respond. Or as Neil Hutchinson, the flight director at Houston, put it, the backup computer "hung up the phone."

Computer engineers compared the problem to a pair of gears that are designed to mesh at a particular point in the operation of a machine. One set is slightly off, so that its teeth do not engage the apertures of the other and vice versa. The fix was to retime the "gears" so that their teeth would mesh.

The launch was scrubbed when it was determined that the reprogramming would take at least a day. Because the propellant tank had to be emptied and then refilled for a second launch attempt, liftoff was postponed 48 hours and rescheduled for 7 A.M. April 12, Palm Sunday.

Sunday morning dawned clear and relatively cool for the ides of April in Florida. Crowds estimated by the Brevard County sheriff to be even greater than those of Friday morning returned to viewing sites and perches. Small craft appeared like schools of fish in the Indian and Banana Rivers, which are not rivers at all but great brackish lagoons separating the mainland of Brevard County from its string of barrier islands. Even the Russian trawler and its Coast Guard monitor were back, 60 miles off shore.

Once again, Young and Crippen rose before dawn and, after the traditional steak-and-eggs breakfast, suited up and rode by van to the launch pad from their quarters in the Operations and Checkout Building in the Kennedy Space Center administration complex.

They rode up in the pad elevator to the 147-foot level, where they disembarked and crossed a narrow bridge formed by the Orbiter Access Arm on the fixed service structure to enter *Columbia*. The hatch was closed, and the crew once more began preparations for launch.

The countdown proceeded smoothly, as it had on Friday. It passed the T minus 20 minutes mark. Hugh Harris, the Kennedy public affairs commentator, announced that the computers were talking to each other. A cheer went up from the throng at the press site.

The late Shorty Powers, voice of Project Mercury, used to ask why it took 500 reporters to cover a single launch. No one ever answered the question to his satisfaction. About 850 were there Sunday, including radio and television commentators, technicians, miscellaneous assistants running about and waving papers, and students. A new generation of feature writers was interviewing some of the gentlemen from the *New York Times* about "the old days." It brought home to the senior journalists that a whole generation had passed since Yuri Gagarin flew an orbit in *Vostok 1* 20 years ago that day.

T minus 9 minutes. There was a built-in hold of 10 minutes. Then the count went down with surprising rapidity. The gaseous oxygen vent arm or "beanie cap" atop the propellant tank was retracted to the service structure. T minus seven minutes. The orbiter access arm was pulled back, leaving the crew committed to the vehicle.

T minus five minutes. The auxiliary power units (APUs) were started to supply power for moving elevons and rudder. T minus four minutes. The main engines were purged preparatory to start.

T minus two minutes, 55 seconds. The liquid oxygen tank was pressurized. T minus 1 minute, 57 seconds. The liquid hydrogen tank was pressurized. T minus 90 seconds. The S-band radar at Kennedy became inoperative, leaving the less reliable C-band radar as the only means of tracking the spaceship at launch until it came within sight of Bermuda station radar. In a quick decision, launch directors at Kennedy and flight directors at Houston decided to trust the C-band to give them all the information they needed.

The count proceeded to T minus 28 seconds. An automatic sequencer took over the remaining functions of the countdown. Launch director George Page read a message from President Reagan: ". . . for all Americans, Nancy and I thank you. May God bless you."

Page said: "Good luck, gentlemen."

"T minus 20 seconds and counting. T minus 15, 14,

13, T minus 10 . . . 4. . . . We have gone for main engine start. . . .

The commentator's voice was squelched by a thundering roar, which rolled over the press site seconds after a blinding gout of orange and yellow flame and vast billows of steam erupted from the base of the Shuttle. The main engines and the solid rocket boosters were firing. The ground shook. A wave of heat washed across three miles of lagoons and swampy fields to the press site.

"We have liftoff!" cried Harris, ". . . of America's first Space Shuttle . . . and the Shuttle has cleared the tower." It was just 3 seconds past 7 A.M. On time.

Columbia rose with astonishing ease and swiftness, not at all like the ponderous Saturn 5 moon rocket, which had seemed to stagger off the pad. The launch had the smartness that solid rocket boosters impart and the smoothness of high-performance liquid-fuel engines.

After clearing the tower, *Columbia* rolled to a head-down attitude. The crew could see the surf falling away. This was the proper attitude for dropping the solid rocket boosters. Mission Control, Houston, came on the line and took over the flight from Launch Control, Kennedy.

"You are Go at throttle up," said the capsule communicator (CapCom or CC), Daniel J. Brandenstein. The voice from Texas came loudly through the public-address speakers at the press site as *Columbia* became a speck in the blue over the Atlantic, trailing a cirrus cloud of vapor.

"Roger," replied John Young.

CC: You're lofting a bit, *Columbia*, so I think you'll probably be slightly high at staging [dropping the SRBs. *Columbia* was flying somewhat higher than its planned trajectory. There was no problem unless an engine failed. Then the lofting would make it more difficult to return to the launch site]. *Columbia*, you're negative seats. [This meant that the spaceship had passed the altitude where the crew could safely use ejection seats in case of engine failure. They would have to stay with the ship.]

At 2 minutes, 20 seconds, Houston announced that the SRBs had been dropped. *Columbia* was on its way to orbit under the million pounds of thrust of its three main engines.

Columbia is launched at 7 A.M. EST April 12, 1981. (NASA)

At 132 seconds into the flight, the boosters are jettisoned. This fuzzy view of the event was recorded by a 16mm automatic camera in the orbiter. (NASA)

Three minutes into the flight. *Columbia* was moving along at 6,200 feet a second (more than a nautical mile a second). It was 51 nautical miles high and 66 miles downrange. Flight director Neil Hutchinson signaled *Columbia* that it was Go for MECO—main engine cutoff.

Still lofting above the planned trajectory, *Columbia* sped northeastward past Bermuda.

"What a view, what a view!" exclaimed Crippen. He had not seen it before. Medical monitors on his body showed his heart rate had reached 130 beats a minute. Young's remained at 80.

"John has been there before," Houston explained.

Six minutes mission elapsed time. *Columbia* was racing at 13,000 feet a second, 76 nautical miles high, 280 miles downrange.

CC: *Columbia,* you are single-engine Rota.
Young: Roger that. [Houston was advising that the spaceship now had sufficient velocity and altitude to make an emergency landing at the Rota Naval Air Station, Spain, in the event two engines failed.]

The message conjured up a vision of a rocket ship leaving Florida at 7 A.M. and landing in southern Spain 18 minutes later. It appears the Shuttle can do that—at a cost of $28 million a flight. Not a cheap commute.

At 6 minutes, 40 seconds, *Columbia* pitched its nose down and began a dive to increase velocity. As it descended from 72 to 63 nautical miles altitude, its velocity rose to 16,400 feet a second.

CC: *Columbia,* you are single-engine press to MECO [which meant the spaceship could get into orbit from that point on one engine].

The spaceship resumed its climb. The main engines were throttled back to 65 percent. At 8 minutes, 34 seconds, Young reported main engine cutoff. *Columbia* was moving at 25,670 feet per second and rapidly gaining altitude.

The big external tank was separated from the spaceship, and the flight computers put the ship through an evasive maneuver to get away from the tank. *Columbia* was in a short-lived preliminary orbit of 81 by 13 nauti-

The camera caught a clearer view of the separation of the external tank 8 minutes and 50 seconds after engine ignition.
(NASA)

cal miles, and so was the tank. In this orbit, the tank would fall back, hopefully into the Indian Ocean as planned, after breaking up in the atmosphere.

Columbia had other business. At 9 minutes, 55 seconds, the crew adjusted the attitude of the ship for the first burn of the twin orbital maneuvering engines. Each put out 6,000 pounds of thrust. In a series of firings over 7 hours, they would complete the insertion of the 100 ton spaceship into a 150-nautical-mile orbit.

Ten minutes, 22 seconds. *Columbia* was climbing to 67 nautical miles and was 1,160 miles downrange. The tank was tagging along at a distance of about 400 feet.

Both orbital maneuvering system (OMS) engines fired for 1 minute, 27 seconds, increasing the spaceship's velocity by 164.7 feet a second. *Columbia* drew away from the tank and took up an oval orbit of 132 by 57 nautical miles.

From launch to the conclusion of the first OMS burn, the crew had been in continuous radio contact with Houston through tracking stations, at Kennedy Space Center and Bermuda.

At the end of the burn, the spaceship passed out of Bermuda's range and sped across the Atlantic toward the Iberian peninsula.

"See you in Madrid," Houston said.

During the interval of silence between Bermuda and the Madrid tracking and data station, controllers had a few minutes of relaxation. The first reusable space transport was in orbit. The initial orbit would have to be raised before it decayed, but it was only preliminary. The propulsion systems had worked better than expected. The orbital maneuvering system performed flawlessly. The first orbital test flight of the new transportation system was looking good.

At launch, *Columbia* had weighed 4,461,620 pounds, with its propellant tank and solid rocket boosters fully loaded. After it shed the boosters and the tank, its weight had dropped to 214,000 pounds.* It was the heaviest vehicle ever flown, even without the 65,000 pounds of cargo it was built to carry.

For comparison: the Apollo command and service modules and the lunar module together had weighed about 100,000 pounds at launch. *Skylab,* including the orbital workshop, Apollo telescope mount, multiple docking adapter, and Apollo, weighed 182,000 pounds.

Columbia was the largest manned vehicle ever flown. Its 122-foot length exceeded that of *Skylab* by 6 feet.

Even before it crossed the rocky coast of Portugal, the spaceship came into signal range of the station outside Madrid. The word from Houston was Go for OMS 2— the second OMS burn.

SC (spacecraft): Okay. We got the targets loaded. How does it look to you?

The crew would be shooting for a 132-nautical-mile circular orbit.

CC: Roger, we see that and it looks good to us.
SC: Dan, that was one fantastic ride. See you at Indy at 36 (at the Indian Ocean station 36 minutes into the flight).
SC: Well, the view hasn't changed any.

It was John Young, speaking from experience. He had made his first spaceflight March 23, 1965, with Gus

Columbia's landing weight April 14, 1981, was reported by NASA as 196,500 pounds. The orbiter structure weighed 156,000 pounds as an empty shell. STS-1 carried no payload but had 9,300 pounds of flight test equipment in the cargo bay.

Grissom in *Gemini 3;* his second July 18, 1966, with Michael Collins in *Gemini 10;* his third May 18–26, 1969, in *Apollo 10* to lunar orbit. On that mission, Tom Stafford and Eugene Cernan descended to within eight miles of the lunar surface in a test of the lunar module. On his fourth spaceflight, April 16–27, 1972, Young commanded the *Apollo 16* expedition to the lunar highlands with Charles M. Duke and Thomas K. Mattingly. STS-1 in *Columbia* was his fifth.

Crippen, a veteran Navy test pilot, had never been in space before and therefore was considered, by the press, at least, a "rookie" astronaut. Space was a new and exciting experience for him, and his elevated heart rate showed it.

"John has been telling me about this for three years," Crippen told Dan Brandenstein at Houston, "but there ain't no way you can describe it. We got a little particulate junk floating around the cabin every now and then and when we get a chance we're going to get the vacuum cleaner and get some of that up."

Young reported that some momentary shaking was felt in the ship when the main engines shut down. It was assumed this was caused by dumping residual fuel. The dumping had been expected to add some velocity to the ship, but Young said he could not tell that it was affecting velocity at all. The venting of residual gases occurred while the OMS engines were firing, but apparently did not add thrust.

John Young at first described the shaking as "pogo," the up-and-down oscillation peculiar to liquid-propellant rocket engines. Devices to suppress pogo had been installed in the orbiter's engines. This was the first chance to see if they worked.

But Young believed the pogo or shaking came not out of the engines but out of the residual fuel dump. "Y'all might look at the data," he said.

South of Spain, *Columbia's* ground track dipped through Gibraltar, Tunisia, Libya, Egypt, Ethiopia and into the Indian Ocean just south of the Seychelles Islands, where contact with Houston was resumed for a few minutes through the Indian Ocean station. Throughout the 36-orbit flight, *Columbia* would pass over southern Europe—Lisbon, Madrid, Naples, and Athens; all of Africa,

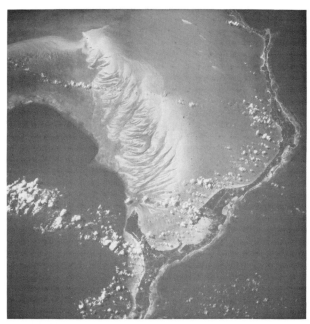

Looking down, the crew got this view of Eleuthera in the Bahamas with the 70mm Hasselblad camera.   (NASA)

Columbia's vertical stabilizer catches the sunlight as the crew flies heads down over the cloud-covered Pacific Ocean.
(NASA)

Below lies Socotra Island off the Horn of Africa. The camera shows clouds and ocean currents driven by winds.   (NASA)

from Cairo to Capetown; Turkey, Lebanon, Israel, Iran, and India, from Calcutta to Sri Lanka; Vietnam (including Hanoi), southern China, near Shanghai; Taiwan (Taipei), Australia (Perth, Melbourne), New Guinea, French Polynesia; and major cities of South America, including Quito, Rio de Janiero, Buenos Aires, and Santiago; Mexico, and the United States, from San Diego to San Francisco and from New Orleans to Philadelphia.

At 44 minutes mission elapsed time, Young fired the OMS engines for 1 minute, 17 seconds, adding 136.5 feet per second to the spaceship's velocity. When the ship came within range of the Western Australia tracking station at Yarragadee, Young reported the burn results. *Columbia*'s orbit was now circular at 132.2 nautical miles altitude. Two more engine burns were scheduled later in the day to raise the ship to its cruising altitude of 150 nautical miles. Before *Columbia* was committed to that height, however, the flight plan called for testing the massive payload doors by opening and closing them. If the doors failed to open, the crew would have to reenter and land on the fifth orbit, because the radiators that dumped excess heat from the orbiter into space were on the insides of the doors. Until the doors opened and the radiators were activated, the spaceship's heat load, generated by electrical equipment and the crewmen, was

View of Cape Cod, Boston, Plymouth, Gloucester, Providence, Taunton, and Worcester.   (NASA)

This remarkable photo was taken as *Columbia* passed over the Himalayas. The Ladakh and Zaskar ranges are etched by snow and shadow.   (NASA)

dissipated by a flash evaporator cooling system. But this used considerable energy and was not designed to operate for a whole mission. Although we think of space as unimaginably cold, the problem in manned spacecraft as far as temperature is concerned is keeping them cool, not warm. Their own electrical systems and crews provide all the heating they need—and more.

Columbia passed between Australia and Tasmania on this first orbit and headed northeastward across the Pacific. Crippen began releasing the door latches—4 on each bulkhead and 16 down the center line where the doors meet. The drill was to open the starboard door first while the port door stayed closed. Then the starboard door would be closed and the port door was opened and closed. This procedure was ordered to make sure that neither door was warped.

The exercise was scheduled for mission elapsed time of 96 minutes, as *Columbia* flew across the United States on its second orbit.*

A word about the payload bay doors. They are the largest U.S. aerospace structures ever made of composite materials, principally graphite epoxy, said to be 23 percent lighter than aluminum honeycomb. Even so, their total weight on the ground is 3,264 pounds. Each door is 60 feet long and 6.7 feet wide, and consists of five segments interconnected by expansion joints. Each is attached to the upper fuselage by 13 Inconel hinges and is opened and closed by electromechanical actuators controlled from the rear flight deck.

Crippen opened the starboard door as *Columbia* passed over Texas between Houston and Ft. Worth. The crew was flying heads down, facing the ground.

"You're missing one fantastic sight," he called. "Here comes the right door and, boy, that is really beautiful out there. Right door now open. We can see a little trash

*Mission Control used the term *orbit,* rather than the older term *revolution,* or *rev,* to describe flight progress. An orbit is a 360° figure in space without reference to the ground location. A revolution is a circuit which returns to the longitude of the launch site. Because of the Earth's rotation, the launch site moved 22.5° of longitude eastward during the 90 minutes it took *Columbia* to fly one orbit. To complete a revolution, the ship would have to fly 22.5° farther in space than it did to complete an orbit. NASA counted an orbit each time *Columbia* crossed the equator on a northeasterly track, or ascending node.

floating out of the payload bay. All the latches work just fine.''

Crippen then closed the starboard door, opened the port door, and closed it. He reopened both doors to expose the radiators. The doors would remain open until the crew prepared to reenter.

"Doors all opened hunky dory," Crippen reported. "We want to show we do have a few tiles missing . . . on both of them [engine pods]." What Crippen saw was also revealed by the onboard television camera. It provided a fine view of the payload bay with the doors open and beyond, discolored gaps where tiles had come off the OMS pods.

For a few moments, observers feared that the nightmare of tiles falling off during launch had become a reality, and rumors flew through the television shacks at the press site that the mission was in trouble.

"My God," a woman cried. "They're going to die!"

Reporters besieged Kennedy Center public affairs people who knew no more than they did, and scrambled for telephones. A calming influence was Bob Crippen's voice coming over the public-address loudspeakers: "The starboard pod has got basically what appears to be three tiles and some smaller pieces off, and off the port pod looks like I see one full square and looks like a few little triangular shapes that are missing."

CC: Roger, Crip. We see that good. [Crippen was pointing the camera lens at the pods.]
SC: From what we can see of both wings, it looks like they're fully intact.

Mission Control counted 15 tiles missing from the orbital maneuvering engine pods. Of all the parts of the ship, the upper rear fuselage was the least susceptible to heat damage during reentry. Maximum heat there was calculated at less than 600° F.

Maxime Faget, a principal designer of the Shuttle, assured everyone that the missing tiles would make no difference at all in the ship's ability to withstand the reentry heat load. The worst that could happen is that a small patch of skin might have to be replaced.

The main heat load would be borne by the underside

Mission Control, Houston was concerned when this view of thermal protection tile loss on the port side orbital maneuvering engine pod was transmitted to Johnson Space Center.
(NASA)

of the ship, and it was there that the thickest and most heat-resistant tiles had been placed. Were any missing there? The crew could not see, and the underside was out of view of the TV camera.

Mission control considered asking the Air Force to photograph the underside of *Columbia* later in the flight with a powerful telescopic camera. A tracking station at Orroral Valley, Australia, relayed a query from Houston asking the crew to describe the color of the OMS pod area where tiles were missing.

SC: Roger. It's red.
CC: Thank you.

There was no further discussion of the subject. The color of the underlying nomex felt insulation was supposed to be gray. Crippen turned to a description of the view over Australia.

"Dan, we can see some big cities down there. See all the lights," he said. *Columbia* was passing just north of Sydney, where it was bedtime.

The spaceship began its third orbit as it crossed the equator over the mid-Pacific. As the vessel passed over the California coast at San Diego, the crew struggled out of their tan pressure suits and made themselves at home. The cabin teleprinter came to life with a message: Troubleshoot the flight instrumentation recorder.

This device kept a tape record of flight instrument readings. The on-off switch had failed earlier and would not shut off the recorder, which was not supposed to run continuously. On advice from Houston, Crippen had shut it off to save tape by pulling a circuit breaker. He might like to tinker with the switch at his convenience, Houston suggested.

"I'll be darned," said John Young. "I just saw Lambert Field down there."

Columbia was passing just north of St. Louis. At 3 hours, 21 minutes, as it was crossing the Atlantic, Houston sent the crew a "Go" to remain in orbit. The crew received it through the tracking station at Dakar.

Two more orbital maneuvering engine burns were coming up to put *Columbia* at its cruising altitude. The third was done at 6 hours, 20 minutes, 45 seconds. The crew fired the right engine for 39 seconds, adding 25.7 feet per second to the ship's velocity.

The shape of the orbit became an ellipse, with the low point (perigee) at 131.7 and the high point (apogee) at 147.5 nautical miles.

CC: It looked like a good burn.
SC: Just had a nice view of the cape coming by. [Columbia had crossed Florida north of the launch site, about over Jacksonville.]

At 7 hours, 5 minutes into the flight, Young fired the left engine for 40 seconds. The boost increased *Columbia*'s velocity by 30 feet a second. It rounded off the orbit at 147.6 by 149.3 nautical miles.

Crippen reported that a transducer was showing high temperature on the right OMS pod. The report was noted, filed, and not referred to again during the flight. It seemed to be typical of the minor flaws that showed up now and then as the mission progressed: the lofting at launch, the faulty recorder switch, the loss of 15 tiles off the engine pods, and a high temperature indication on one pod.

None of these conditions was regarded as serious. What impressed Houston was the accuracy with which the orbit maneuvers had been carried out. *Columbia*'s overall performance was magnificent.

Now in cruising orbit, John Young was moved to make one of his rare pronouncements—to which he was entitled as commander.

"The flight has gone as smoothly as it could possibly go," he said. "We have done every test that we are supposed to do and we are up on the time line and the vehicle has just been performing beautifully, much better than anyone ever expected it to do on the first flight, and no systems are out of shape. We did three star tracker alignments in less time than it takes to do one in the mission simulator. All the RCS [reaction control system] jets have been fired. The vehicle is performing like a champ."

At 10 hours, 52 minutes elapsed time, there was supposed to be a private medical conference between the crew and a flight surgeon at Houston. Neither Young nor Crippen had any complaints, and the session was scrubbed.

Houston advised the crew that missing tiles would not present any problem. Another advisory: A low-energy

Flight Commander John W. Young, having settled down to orbital flight in a brand new spaceship, turns toward the camera his pilot, Robert L. Crippen, is holding. This is a view of the commander's seat on the flight deck. A loose-leaf flight data notebook floats conveniently beside the command station in the microgravity of free fall around the Earth.   (NASA)

peak of a solar flare which had been detected April 10 was due to reach Earth at 11 hours into the mission.

"No concern about radiation effects," the capsule communicator said. "However, we do want to caution you that during the next four revs as you pass through the South Atlantic anomaly,* there may be enough radiation to kick off your smoke detector."

"Good grief, Henry!" exclaimed Crippen, addressing the communicator, Hank Hartsfield. "That sounds like fun. Who thought of that?"

CC: Also there's a good chance there's going to be a spectacular aurora later this evening.

Houston urged the crew to take photographs if they saw an aurora.

Now and then, the dialogue with Mission Control exhibited more sentiment than the usual test pilot report. The crew paid a tribute to John Bjornstad and Forrest Cole, the launch pad technicians who perished from nitrogen asphyxiation.

"I'm sure they would be thrilled to see where we have the vehicle now," said John Young solemnly.

The first day was ending. At 8 P.M. eastern standard time, at 13 hours elapsed time, a sleep period was scheduled. Instead of the fairly snug wall hammocks depicted in orbiter plans, the crew was required to sleep in their ejection seats this trip. They had to be ready for anything.

Communication ceased. Houston announced that *Columbia* was settled in an orbit of 146.3 by 149.3 nautical miles with a period of 89 minutes, 54.8 seconds and an inclination of 40.2° to the equator. Cabin pressure, a normal mixture of oxygen and nitrogen, was 14.7 pounds per square inch—the same as at sea level.

Columbia was the first American manned spacecraft to provide a normal breathing atmosphere for the crew. Mercury, Gemini, and Apollo atmospheres were pure oxygen at about five pounds pressure.

On the flight deck at sleep time, the temperature was 79.2°F. But it dropped during the sleep period.

*A region where the Van Allen radiation zone dips to low altitude.

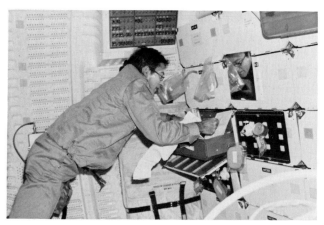

On the mid-deck, Young cleans off his razor after shaving. At the right is a food tray, with a sandwich and processed food in a can. The food containers seem to be pointing sideways, but float in place. (NASA)

At 4:30 A.M. EST, April 13, Young and Crippen awoke with the complaint that the deck had become too cold for comfort. Houston sent up a wake-up call in the form of a raucous country and western style ballad about the Shuttle, which advised the crew to "just lay back and let 'er fly."

Once awake, and no one could have remained asleep through that rendition, Crippen reported that the temperature had dropped to 76°F., or so it said on the console instruments. That did not seem chilly to Houston, but controllers advised that if the crew wanted it warmer, they could simply turn down the cooling system.

An hour later, at 22 hours, 30 minutes elapsed time, Young put *Columbia* into a gravity gradient attitude. This is a maneuver first developed in the Gemini program.

Nose down, so that it was approximately perpendicular to the surface of the Earth, the spaceship would eventually become nearly motionless, like a plumb bob. It would be anchored in space—to the Earth's center of gravity—and thus conserve fuel which otherwise would have to be expended through the thrusters to keep it stable.

Over a fairly long period of time, stability in gravity gradient attitude is achieved by the fact that gravitational force is stronger on the part of the ship that is nearest the

Earth than on parts farther way. Eventually, the force acting on the nearest part—in this case the nose of *Columbia*—overcomes inertial forces acting on the ship and damps them out. Rolling, yawing, and pitching motions die away.

Young and Crippen let *Columbia* drift in this position for 3 hours and 15 minutes. Over Australia on the 16th orbit, they played a tape recording of "Waltzing Matilda" to serenade the night watch at the Orroral Valley station.

Over the sunlit side of the Earth, they rotated the ship slowly, barbecue style, to measure solar heating on the underside and top side of the wings and fuselage.

It would have been a sight to see—a spaceship, with wings, the size of a DC-9 commercial airliner, moving around the Earth at five miles a second while anchored nose down, tail up to the planet's center of gravity. Yet Young and Crippen had no sensation of gravity. They seemed weightless. Physically, they were falling around the Earth in *Columbia,* and free fall creates the illusion of zero gravity. Nevertheless, the gravitational force field gripped the spaceship firmly; held it in an orbit defined by its velocity. Until the balance between the ship's kinetic energy and gravity were altered, *Columbia* would stay in its celestial track. Retrofire by the space engines would change the balance, and *Columbia* would come down to Earth.

At 23 hours, 53 minutes mission elapsed time, as *Columbia* crossed the west coast of Mexico on its 17th orbit, the crew learned the results of its "Waltzing Matilda" serenade at Australia.

CC: Orroral Valley says thanks for the hometown music.

Conversation resumed, first through the station at Tula Peak, New Mexico, and later through MILA, the station at Kennedy Space Center.*

CC: *Columbia,* if you'll look down, you'll see Cape Kennedy, perhaps. There was a tremendous launch from there yesterday which you may not have seen.

*MILA is the acronym for Merritt Island Launch Area, the original name of the Kennedy Space Center.

Crippen performs some acrobatic rolls and floats a bit.

(NASA)

SC: Oh, we saw it. I got the runway and the VAB [Vehicle Assembly Building] in sight.

Young and Crippen sent television pictures of their activities on the flight deck down to Kennedy. Some of the images were quite clear. The crewmen floated and somersaulted about and drifted up and down the ladder between the flight deck and the mid-deck.

The northeasterly course allowed the camera to show the east coast from Jacksonville to Cape Hatteras, looking through the open payload bay.

When *Columbia* came within range of Madrid, the crew had good news. The cabin temperature had reached a comfortable 79°F.

During the morning of the second day in space, after the gravity gradient experiment ended, the crew carried out a sequence of attitude changes with the reaction control thrusters. The effect of the exercise was to reshape the orbit to 144 nautical miles, circular.

Crippen reported another minor glitch—a leaking oxygen regulator valve. Mission Control checked it and advised not to worry as long as the regulator continued working.

Crippen was fascinated by the ease of moving around in free fall. "I tell you, it's going to be tough to go back to work in the 1-G trainer after finding out how really easy it is to get around in this vehicle."

CC: I think you may be spoiled now.

It was still chilly on the flight deck, the crew complained. They sounded like tenants asking the landlord to turn up the heat. Finally, Houston advised; "John and Crip—if you get extra time you might check under the mid-deck floor. That heat exchanger valve should be in the full hot position. We are a little puzzled over some of the thermal data we've got down here and the fact it's getting cold."

SC: The same condition we had last night. The valve was in full hot and we're still freezing. John's putting on his long underwear right now.

CC: Crip, did you all look at that valve? Under the mid-deck floor? We think it may not be pinned hot and the automatic system might be changing it on you. Pin it full hot and we'll track it. If that valve does not work, just turn the water loop one on.

SC: Wait a minute! John says it looks like it's on full cold now.

CC: Ahah! Okay!

As *Columbia* soared across the California coast on the 20th orbit, the Buckhorn tracking station near Edwards passed up a weather forecast: "Okay, John. The winds forecast for Edwards tomorrow—about 10 knots out of 240 degrees.

SC: Sounds good.

The crew continued firing reaction control thrusters to check performance, fore and aft. Thrusters would be fired to help stabilize the spaceship's glide downward through the high atmosphere to supplement the aerodynamic control surfaces.

"You can certainly hear these big thrusters going off up there in the nose. They really move the vehicle. It's really sporty," Crippen said.

The cabin was beginning to warm up. At mid-morning, Young reported that a thrust monitoring instrument was not operating properly. It was not reading the correct velocity.

Time out for lunch. Crippen reported he was enjoying a corned beef sandwich, courtesy of John Young. That delicacy was not on the menu, which featured mainly dehydrated foods elaborately and expensively processed and packaged by dieticians.

Mission Control responded to the contraband sandwich with pretended shock. "My, oh my," said the capsule communicator.

Further comment was forestalled by a telephone call from Washington. Vice President George Bush was on the line, waiting to talk to the astronauts.

Young: Hello, Mr. Vice President. Yes, sir, we're just having a lot of fun up here.

Bush: Hey, listen. I'm glad to talk to both you and Crip. How's he behaving?

Crippen: I'm trying to behave pretty well, Mr. Vice President.

Bush: Well, Crip, this is pretty far away from when we were doing our running down there [at the Space Center],* and I've just come from seeing the president. How's it going up there? Everything all right?

Young: The spaceship is just performing beautifully.

Bush: Well, that's great. I think that your trip is just going to ignite excitement and forward thinking for this country, so I really just wanted to call up and wish you the very best.

Columbia cruised. Its second day in space was dedicated to testing the attitude control system. Young and Crippen were well satisfied. They yawed, rolled, and pitched the ship up and down. They stood it on its head and on its tail. Houston saw most of the results on telemetry. *Columbia* was a high-performance spaceship.

Bedtime. Shortly after Houston said good night, an alarm sounded on the flight deck. Young and Crippen came on full alert. They found that it had been triggered by a sudden drop in temperature in one of the three auxiliary power units. Two of the APUs had to be operational to provide power for the aerodynamic flight controls after reentry. A cold power unit might be hard to start. It looked like the onset of a hairy situation.

Houston advised the crew to recycle the heating system switch to engage the A heater. Young and Crippen did so and returned to sleep, but the problem did not go away.

*Bush referred to some jogging they did during his visit to Kennedy.

They were awakened shortly after 3 A.M. EST and cooked breakfast on the hot plate.

SC: John even had the coffee made when I got up this morning.

CC: Well, it looks like you're going to have a pretty day for entry. Switch the APU gas heater to B so we can watch it. [They had left the heater on A during the night.]

SC: We got a nice view of Italy this morning. There's an old favorite . . . Naples.

It was John Young speaking. *Columbia* was in its 31st orbit, with 5 more to go before deorbit—the next big test. The cold power unit began to worry Mission Control.

Sensing a potential crisis, news people pressured public affairs spokespeople for a prognosis. What was the worst possible case? Public Affairs hastened to reassure the media that "confidence is high it will be possible to start the malfunctioning APU." That made the situation sound ominous.

As mentioned earlier, the APUs are 138-horsepower turbine engines fueled by gaseous hydrazine. They drive the hydraulic actuators that move the aerosurfaces: the elevons, rudder, speed brakes, and body flap. They also provide power so that the pilot can lower the landing gear on final approach.

The APUs do not run all the time. They are started five minutes before launch and shut down, with the main engines, at orbit. During orbital flight, they are inactive. They are kept warm to guarantee flow of fuel and lubricant by redundant heaters (A and B) which are supposed to prevent temperature in the units from falling below 45°F.

Five minutes before the deorbit burn, the APUs are restarted. All three are supposed to be operative, although two can control the descent. If one unit became too cold to start, the crew could come back on two—but if one more failed, control would be degraded.

There was still plenty of thermal margin for APU 2, the one cooling down. On orbit 32, its temperature was 186°F. However, it had dropped from 262°F. on orbit 31.

As long as the temperature did not fall below 80°F, there was nothing to worry about, Public Affairs said. The crew discussed the prospects with Houston.

SC: Am I to understand right now that we anticipate no modification of APU start procedures?

CC: It is probable that you will have to use start override to get it started, but otherwise no problem at all. And we have test data that show an APU starting with temps much colder than what you will be carrying at the time you start them.

SC: Okay. But do I understand you want me to try a normal start initially?

CC: Bob, the plan now is we want you to start override the first try, even. That means you will have to lead the start by three or three and a half minutes.

Flying across the Pacific on orbit 34, the crew reported picking up a signal from the St. Petersburg, Florida, TACAN far beyond its normal range.* Radar showed *Columbia*'s orbit had settled down a bit, to 143 by 147 nautical miles. Thruster firings probably accounted for this, although some minute atmospheric effect might have been present.

As *Columbia* passed over the United States, Houston sent a "Go" for the crew to switch the flight computers to the reentry program.

Deorbit time on the 36th orbit was scheduled at mission elapsed time of 2 days, 6 hours, 30 minutes. Young and Crippen began to struggle into their tan pressure suits. They prepared to close the payload doors and maneuver the ship to deorbit attitude.

Orbit 35. "You just crossed the coastline," said the capsule communicator. "I guess northern California."

SC: See the Great Salt Lake and Ellington Air Force Base. [This was quite an expanse—Utah to Texas.]

SC: Looks like we are going right over Chicago. Or are we lost in space again?

CC: No. We think you're probably right. [The ground track passed south of Chicago. The metropolitan complex of northeastern Illinois and the Indiana Calumet region along Lake Michigan were off to the left.]

*A Tactical Air Navigation beacon—one of several *Columbia* used to check its course.

Two hours and 15 minutes from deorbit. The crew took time out for a snack. Interrupted conversation continued through Bermuda, Dakar, Ascension Island, and Botswana stations.

CC: A reminder. Go to Start override at six and one-half minutes.

SC: Roger, Joe. We are maneuvering to IMU alignment.

The IMU is the inertial measuring unit, a navigation instrument which measures every motion of the vessel and thus determines where it is in relation to a fixed point. Houston reported that an analysis of the IMU alignment showed that *Columbia*'s maneuvers were precise and that the spaceship was exactly where it was supposed to be at "every point in time."

At mission elapsed time 51 hours, the crew received a call from Houston through Hawaii asking them to push in the flight instrument recorder circuit breaker so that it would be running during the descent.

As *Columbia* crossed the California coast, an Air Force optical tracking telescope at Anderson Peak, Big Sur, picked it up and transmitted the image to Houston by television. In years gone by, the telescope had been used to photograph Apollo flights.

The sun was hot now on the desert, and the margins of the lake bed were black with people and vehicles. A six-mile traffic jam was reported on all roads leading to Edwards Air Force Base.

Orbit 35 merged into orbit 36 over the western Pacific. *Columbia* crossed the equator on the last full orbit of the mission at an elapsed time of 52 hours, 20 minutes. Astronaut Joe Allen, the capsule communicator, came on the line with a message as it crossed America.

"John and Crip. We are 30 seconds from loss of signal. Want to report *Columbia* is in super shape. Almost no write-ups. We want her back in the hangar and you have a preliminary "Go" for the burn."

SC: Okay, Joe. We concur. Flying like a champ.

Orbit 36 took *Columbia* down the Atlantic missile range. When it passed out of signal range of MILA and

Bermuda, its next contact was the station on Ascension Island between Brazil and West Africa.

Ascension coming up, the public affairs officer said. John Young reported that he was standing by to open the auxiliary power unit tank valves preparatory to starting them.

CC: Okay. We're watching [on telemetry]. Go ahead. APUs look good to us. You can start to maneuver to burn attitude whenever it is convenient.

Young signalled the computers to turn the ship around so that it was flying tail first with orbital maneuvering engines pointing in the direction of flight, slightly above the horizon.

CC: *Columbia*, your burn attitude looks good to us. Everything aboard looks good to us. You are Go for deorbit burn.

SC: Okay. We understand. Ready for deorbit burn. Thank you, now. That's the best news we've had in two and a half days. About looking good. I wonder if you were including the crew.

CC: There are exceptions to every rule.

Loss of signal. A few minutes of silence. Communications were restored through Botswana.

CC: Do a good one. We will see you at Guam.

SC: Okay.

CC: Sorry. Make that Yarragadee in about 15 minutes.

SC: You have so many stations you just can't make up your mind what one we are at.

CC: Roger that.

The deorbit burn would take place over the Indian Ocean, but Houston would not know how it came out until *Columbia* came within range of the Australian tracking station at Yarragadee.

At mission elapsed time of 2 days, 5 hours, 21 minutes, 30 seconds precisely, the maneuvering engines fired for 2 minutes, 39.5 seconds. *Columbia*'s velocity was reduced by 297.6 feet per second. Its orbit changed from a circle to an ellipse that intersected the ground. Apogee was 146 nautical miles, perigee zero.

Young turned the ship around so that it was flying nose first to prepare for interface—contact with the "sensible" atmosphere. Interface would come in 27.5 minutes at an altitude of 400,000 feet, to the east of Australia. Interface was the first moment of reentry. From there, *Columbia* would become a heavy glider and descend along a path of 4,400 miles through the atmosphere to California.

Spaceships had been launched before, had flown in orbit before, had reentered before; but nothing like that long glide across the Pacific Ocean had ever been attempted.

The mission control room at Johnson Space Center, Houston, was filled with people breathing deeply, and the multitudes standing out on the desert floor at Edwards waited restlessly.

CC: *Columbia,* this is Houston, through Yarragadee. We are standing by.

A crackle of static, a hum, and then:

SC: Burn was on time and nominal.

At Houston there were sighs, at Edwards cheers and applause. *Columbia* was coming home.

SC: APU 3 started fine also. We have got 2 and 3 running now.

CC: Okay, John. We copied the shortest of all burn reports. And Crip . . . We understand that you have two of three APUs running now.

SC: You understand correctly, but you're fading now.

The third power unit was not to be powered up until entry interface, which was coming up shortly.

CC: . . . You will like to know that four chase aircraft have just been launched from Eddy and are coming up looking for you.

SC: Yeah. We ought to be there in about 45 minutes.

At Guam, Houston saw the third power unit start up on telemetry.

SC: We're showing about 85 miles [altitude] and we're about 0.45 or 0.442 [gravities]. The atmosphere was slowing *Columbia* down. But interface was still 90 seconds off].

CC: Rog. Moving right along. Nice and easy does it, John. We're all riding with you.

Columbia crossed the equator and began its 37th orbit, which would end in California.

At Edwards, crowds surged up to a mile-long security rope Air Force guards had strung along the lake bed. It was 10 A.M. pacific standard time. The sun was high and hot. Air shimmered out on the dry lake bed to create the illusion of water.

Columbia entered a 16-minute radio blackout as temperatures on its heat shield rose to 2,750° F. at the wing leading edge. The heating ionized the air around the ship so that it was enveloped in an electrically charged sheath that blocked radio signals. This is the blackout effect. As the ship cooled, the effect dissipated.

Two C-band radar contacts with the ship were reported by the Edwards public-address system. *Columbia* was descending through 185,000 feet, 410 nautical miles out. The crowd roared with delight.

SC: Hello, Houston. *Columbia* here.
CC: Hello, *Columbia.* Houston here. How do you read?
SC: Loud and clear. We're doing Mach 10.3 at 180 [thousand feet].
CC: *Columbia,* you've got perfect energy. Perfect ground track. John, we show you rolling right. Looking good. We show you out of 151,000 now. Mach 8.4. Looking good.
SC: Roger that. I got a solid TACAN lock on TACANs 2 and 3.
CC: All three APUs are hanging in there. Looking good. We show you crossing the coast now. *Columbia,* you're out of 130,000, now Mach 6.4, looking good.
CC: 112,000. 4.8 Mach.
Public affairs officer: Range is 130 miles. John Young is rolling [making a sequence of S-turns to reduce velocity. He was flying *Columbia*—through the computers, but now the commander was telling them what to do].
CC: Seventy thousand feet, Mach 1.3. Forty-two miles. *Columbia,* we show you slightly high in altitude. Coming down nicely.

Public Affairs had predicted a mild sonic boom as *Columbia* zoomed by and then made a turn into final approach. Boom! Boom! A double cannon shot startled the crowds. *Columbia* swept high across the dry lake and the hills east of Edwards. Young rolled the big glider into a wide, looping turn to the north and back west, around Leuhman Ridge and the town of Boron. He headed down toward the lake and the seven-mile runway. Thousands

peered into the pale blue sky for the first glimpse of the ship.

CC: *Columbia,* you're subsonic now, out of 50,000. Turning on final. Your winds on the surface are calm.
SC: That's my kind of wind.
CC: You're right on the glide slope. Sixteen thousand feet, 271 knots. Eleven thousand feet.

On April 14, *Columbia* was due to land on the dry lake bed runway at Edwards Air Force Base, California. Wheels are down as the spaceship glides swiftly toward the runway.   (NASA)

Columbia has stopped. The first orbital test of the Shuttle has been successful and a new era in manned spaceflight begins.   (NASA)

Contact. *Columbia*'s main landing gear is on the ground and the nose gear is coming down.   (NASA)

Crippen descends from *Columbia,* following Young, who stands at the foot of the stairs. Both wore pressure suits for landing as well as for launch.   (NASA)

Houston continued to count *Columbia* down: nine thousand, five thousand, twenty-five hundred. Landing gear down.

Columbia suddenly appeared over the lake like a fat bird coming home to roost. The clamor of the crowds overrode the big glider's mild wash of air as *Columbia* flared and settled down on the runway. The main landing gear tires hit the dirt at 215 knots. The nose wheel came down seconds later. *Columbia* rolled 8,993 feet to a stop. It was 10:20:52 A.M PST.

CC: Welcome home, *Columbia*. Beautiful!
SC: Do I have to take it up to the hangar, Joe?
CC: We're going to dust it off first.
SC: This is the world's greatest flying machine, I'll tell you that. It worked super.

Columbia is serviced at the Dryden Flight Research Center, Edwards. Refrigeration and purging trucks are connected by long hoses to the ship's cooling system and gaseous purge duct. (NASA)

First Flight of a Used Spaceship

The concept of a winged spaceship that went up like a rocket and came down like a glider had been proved. Now, the second design goal of the Shuttle—its reusability—remained to be tested.

Columbia was airlifted back to the launch site on its Boeing 747 carrier, arriving April 28 to be refurbished and refueled for its second spaceflight. NASA headquarters announced that the second orbital test flight would be launched September 30. It would be a five-day mission with a cargo of scientific instruments and a 50-foot arm that could lift satellites in and out of the payload bay during orbital flight.

By early June, it was evident that the September 30 launch date was overly optimistic. Refitting *Columbia* for another mission was taking longer than expected. At the outset of Shuttle development, Congress was told that the turnaround time between landing and launch would be two weeks. George Page, the launch director at Kennedy, and his launch crew were finding that the job was

taking six months—at least the first time. Undoubtedly, the turnaround time could be reduced later, with experience. Page and his crew doubted that turnaround could be done in less than two months, at least during the test program.

Meanwhile, a preliminary flight review at the end of April produced some good news and some bad. No design problems appeared, but there were unexpected effects. *Columbia* tended to fly 5 percent higher than its planned trajectory. This was attributed to excess thrust of the solid rocket boosters (SRBs). The steeper climb made it more difficult for the crew to return to Kennedy in the event of main engine failure.[1]

Heat shield damage was more extensive during the ascent than expected. The blimplike propellant tank shed pieces of ice onto the tiles, chipping and scoring more than 300, mainly those on the underside of the orbiter, where heating was highest during reentry and descent. The tiles shaken off the orbital maneuvering engine pods

Columbia is rolled to the mate-demate device to be prepared for the ferry flight back to Florida. The work continues through the night. (NASA)

Columbia returns to Kennedy Space Center aboard the 747 transport April 21, 1981, wearing tail cone after its initial shakedown voyage. (NASA)

pointed up a weakness in the shielding. The pods were retiled carefully.

Heat loads during reentry and descent were 400° F. less than expected, however. And there was no tile damage from heating. *Columbia* had slipped easily into the atmosphere, proving that a winged vehicle could do so safely. Despite installation difficulties, the heat shield worked, and 99 percent of it had stayed in place.

Although minor failures and glitches plagued the flight, they did so to a lesser degree than on any first test of a manned spacecraft that I have observed—and I have seen all of ours. The main problem was the loss of the flight recorder as the result of a defective switch. Critical pressure and temperature data were missed during the descent. The loss influenced an important decision that contributed to reducing the length of the second mission.

Intermittent problems with an auxiliary power unit heater and a short circuit in the instrumentation of the steering thrusters appeared and were resolved. The only complaint Young and Crippen mentioned later was the failure of the toilet. It stopped working during the flight.

"We did not see anything of a fundamental design nature that would cause problems," said Robert F. Thompson, the Shuttle program manager.

The most serious problem of the flight seemed to be the last one to be disclosed to the public. Films of the liftoff revealed that the backwash of solid rocket booster exhaust during the first few seconds of launch shook the orbiter beyond its safe structural limit. The backwash created a pressure wave of two pounds per square inch, four times the pressure anticipated. It moved the wing elevons six to eight inches despite the fact they were locked in, and also moved the body flap on the underside of the orbiter. It bent a forward strut.

This problem had to be dealt with before the next flight. Otherwise the cargo of delicate scientific instruments and the manipulator arm would suffer damage, even if the orbiter structure did not.

Although *Columbia*'s flight performance was generally satisfactory, it could not meet the design goal of lifting 65,000 pounds to low Earth orbit on an eastward launch or 32,000 pounds on a polar launch from Van-

denberg Air Force Base, California. The best *Columbia* could do was lift 37,400 pounds from Florida on the easterly launch and 22,000 pounds on the polar launch.

Even with a 9 percent increase in main engine thrust and a reduction in the weight of the SRBs, *Columbia*'s cargo capacity could be raised only to 57,000 pounds on easterly launches, which took advantage of the Earth's rotation. The cargo weight limitation would constrain the heavier military payloads destined for polar orbits.

However, the major factor reducing the utility of the Shuttle as a transport was the longer-than-planned turnaround time. This had become painfully apparent during the slow refurbishment and inspection process over the summer of 1981.

OSTA-1

The second flight was a historic event, inasmuch as it marked the first time a space vehicle had been used again. Beyond this, however, it would demonstrate, hopefully, the utility of the orbiter as a platform for scientific instruments that could observe the space environment and monitor the Earth's surface from orbit.

In the forward section of *Columbia*'s 60-foot cargo bay, technicians installed a U-shaped pallet 9.5 feet long on which five instruments were mounted. The pallet was an engineering model of a unit in the European Space Agency's Spacelab system. It was built and flown to Kennedy by British Aerospace for ESA.

The instruments, consisting of cameras, radiometers, and other sensing equipment, represented a demonstration by NASA's Office of Space and Terrestrial Applications (OSTA) of remote Earth-sensing technology.

The instruments were designed to photograph and identify the nature of geological formations where hydrocarbon and mineral wealth might be found; the nutrient regions of the sea where fish might be found, and the incidence of carbon monoxide in the atmosphere of the northern and southern hemispheres. Carbon monoxide is partly a man-made pollutant, produced by slash-and-burn agriculture, burning of fossil fuels, and internal combustion engines. In addition, the crew was to take

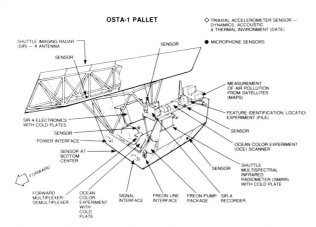

On its second mission (STS-2) *Columbia* carried the first scientific payload of the Shuttle program in the cargo bay for NASA's Office of Space and Terrestrial Applications (OSTA). Mounted on a British Aerospace pallet, an element of Spacelab, was an array of experiments including the amazing Shuttle Imaging Radar—A (SIR-A), a camera so powerful it could photograph a prehistoric landscape buried beneath the sands of the Sahara (described later in this chapter).   (NASA)

Crippen (left) and Young brief Joe H. Engle (right) and Richard H. Truly, the prime crew for STS-2, at Johnson Space Center, Houston.   (NASA)

pictures of lightning strokes as the spaceship flew over thunderstorms.

The principal cargo bay test, however, was that of the remote manipulator system (RMS), a 50-foot-long arm with shoulder, elbow, and wrist joints and a grasping "hand." It was developed by Canada and was the boom or derrick the spaceship would use for deploying or retrieving satellites in orbit.

Because it was fashioned after the human arm, the RMS became generally known as the "arm." It could be operated by the ship's computer or by a crew member at the rear deck console. NASA said the arm had cost Canada $100 million to develop. This one was a contribution to the Space Shuttle Transportation System, but the space agency would purchase additional arms when needed for additional orbiters.

The crew for STS-2 consisted of two astronauts we met before—Colonel Joseph H. Engle of the Air Force and Captain Richard H. Truly of the Navy. Engle, now 49, was looking forward to his first spaceflight, although he had flown to what the Air Force once called the "edge of space" as an X-15 pilot. Truly would pass his 44th

birthday on the flight—his first space venture. Both had flown the *Enterprise* on approach-and-landing tests in 1977.

During most of the planned 124-hour mission, *Columbia,* with cargo bay doors wide open, would fly "upside down," so that the instruments and the crew's heads looked toward the ground. In orbital free fall, the crew would have the impression of looking up at the Earth, and that is the illusion that was created by television pictures of the cargo bay showing the Earth above and the heavens below the spaceship.

In order to use the cameras and radiometers to maximum advantage, the altitude of the flight was reduced from the 150 nautical miles of STS-1 to a 140 nautical or 161 statute mile cruising orbit. The flight was to be inclined 38° to the equator, instead of 40° as on STS-1, covering the planet from 38° N. to 38° S. latitude.

When the Shuttle test flight program was laid out in 1976, the planners figured that the orbiter's cargo bay would face the Earth about 88 hours out of 124. This attitude made it feasible to use the orbiter as an Earth observation platform.

Along with the pallet, the experiments weighed 4,925 pounds and took 1.4 kilowatts of power from *Colum-*

bia's fuel cell batteries. Each of the five instruments was designed to collect data in a different mode.

One of the most promising of them was the newly developed radar camera—the Shuttle Imaging Radar-A, or SIR-A. It could take images of the ground in clear skies, through clouds, or at night. Radar provides its own illumination.

This camera was an improved version of a synthetic aperture radar imaging system first tested successfully in 1978 on the short-lived *Seasat* orbital satellite. Although shut down early by a short circuit, the *Seasat* camera radioed to Earth remarkably clear and detailed photographs of terrain in the Pacific Northwest. The instrument was designed for ocean surveillance, however, where surfaces are mostly flat, and its microwave beam was projected almost straight down, to be reflected almost straight back.

The Shuttle camera was a side-looking radar instrument that sent forth a microwave bean at an angle. This enabled the reflected energy to show relief—slopes, hills, mountains, valleys—in considerable detail.

The "lens" of the camera consisted of an antenna 30 feet long and 7 feet wide. It collected microwave energy reflected from a ground target as a photographic lens collects light waves. By mounting the antenna on a moving platform and emitting a succession of radar energy pulses, the aperture of the antenna lens is synthesized, or extended, from 30 feet to 8 miles.

The Shuttle camera actually was built from spare parts of the *Seasat* instrument and used the same, L-band (23-centimeters), wavelength. The transmitter put out a kilowatt of microwave energy. When the beam hit the ground, its energy density was diffused to the level of a commercial television station signal.

Experimental use of SIR-A could, if successful, set off a revolution in the technology of orbital reconnaissance. Clouds or nighttime didn't matter. This camera could show what was on the ground any time, with a resolution of 36 meters—about the size of a football stadium.

If it worked as expected on *Columbia*, the Jet Propulsion Laboratory team managing the experiment would put a similar camera on a Venus-orbiting satellite to photograph the hidden landscape of Venus through its

dense veil of clouds. But this project, the Venus Orbiting Imaging Radar (VOIR), was not funded.*

Aboard *Columbia* the camera was supplemented by another instrument, a Multispectral Infrared Radiometer. The receiver picks up wavelengths in the electromagnetic spectrum that reveal the presence of certain minerals.

Radiometers aboard the Landsat satellites have picked up iron oxide stains on rocks, thus enabling the U.S. Geological Survey to map iron oxide deposits in the American West. The Shuttle radiometer was designed to detect other spectral signatures indicating the presence of clays and their constituent minerals in the surface material.

The radiometer would help identify the chemical nature of surfaces photographed by the radar camera. The two promised big advances in geological reconnaissance from orbit.

A third experiment, called FILE (Feature Identification and Landmark Experiment), was designed to automate reconnaissance from orbit. Its sensors could identify such features as vegetation, rocks, soils, water, snow, and clouds. The instrument could be programmed to take a certain number of photographs of each feature and skip repetitive scenes.

The selectivity of FILE would reduce analysis time, an important consideration for military surveillance of the surface as well as for scientific and commercial observations. For example, such an instrument could be programmed to record photographs and other information from 30 wheat-growing areas and ignore data from other cultivated regions.

A fourth experiment on the pallet surveyed global concentrations of carbon monoxide in the atmosphere, for the first time from orbit. It was operated by an optical detector, a gas filter radiometer, a camera, and a tape recorder. A fifth instrument was a super fish finder—an eight-channel ocean scanner. It could detect the green

---

*The JPL proposed a cheaper version of VOIR in 1982–1983, called the Venus Radar Mapper. It would produce landscape scenes at poorer resolution than VOIR but, if successful, produce radar photo maps of Venus.

color of chlorophyll in marine algae as an indication of fish populations in an ocean area.

In the cabin, a dwarf sunflower in a container was the pioneer botanical experiment aboard the Shuttle. It was simply stowed in a locker during the flight. The plant was to be the subject of a more drawn-out experiment on *Spacelab 1*. Its purpose on STS-2 was to indicate the moisture requirement of the plant during free fall.

OSTA-1 did not seem dramatic, but the overall demonstration of the orbiter cargo bay as a scientific instrument platform in orbit would fulfill a major Shuttle objective.

A NASA program scientist for the OSTA experiments, James V. Taranik, predicted that the Shuttle ultimately would change the way Earth-related research is done in space.

The Debacle of October 9

During July, the Shuttle directorate made public its concern about the potential damage of the backwash, or overpressure, of solid rocket booster exhaust at liftoff. Although STS-1 had escaped with minor damage, the payload of scientific instruments on STS-2 and the arm were vulnerable to shaking.

Kennedy and Marshall engineers decided to install a powerful water spray to dampen the shock wave. It would deluge the flame pits with 70,000 gallons of water a minute. Model tests at Marshall seemed to verify the idea's feasibility. The spray was installed on Pad 39-A at a cost of $1.5 million, and a test September 16 while *Columbia* was on the pad indicated it would work, although the rocket engines were silent.

A revamping of the prelaunch work schedule as a result of the spray installation caused the postponement of the launch date from September 30 to October 9. The announcement seemed to trigger a reactivation of Murphy's Law.

On September 22, three gallons of caustic nitrogen tetroxide were spilled over the orbiter's nose as the oxidizer tank of the forward steering thrusters was being filled. The nitrogen textroxide reacts with hydrazine fuel to produce thrust on contact.

The effect of the spill was to loosen 378 heat tiles on the fuselage by dissolving the adhesive that binds them to the vehicle. The highly reactive oxidizer also contaminated the thruster engine housing.

The initial damage assessment indicated that if *Columbia* had to be taken down from the pad and rolled back to the Vehicle Assembly Building for cleanup and tile replacement, the launch might be delayed until December.

Investigation showed that the spill was caused by a defective valve. It failed to close when the shutoff was operated from the ground supply. When the valve was examined, it was found that it had been contaminated by deposits of iron nitrate, apparently the product of a reaction of the oxidizer and its steel container.

"We never considered this type of failure probable," George Page, the launch director, told a news conference. "Murphy's Law sort of got us there."

Rockwell engineers advised that the reaction control engine housing could be entered by removing a panel, and thus the housing could be cleaned up without removing *Columbia* from the pad. Scaffolding was erected around the orbiter, and crews went to work like steeplejacks to decontaminate the housing and replace the tiles.

In mid-October, NASA was ready once more to launch *Columbia*, on its second orbital test flight. Liftoff was set for 7:30 A.M. November 4.

Scrub!

Once more, the beaches, causeways, and approaches of the huge Kennedy Space Center swarmed with people and vehicles. The crowds were estimated by the sheriff at 600,000. Automobiles, vans, recreational vehicles, pickup trucks, motorcycles, scooters, and bicycles were parked over square miles of grass and sand.

The press and VIP throngs were smaller than before. Still, busload after busload of political and entertainment personalities rolled into the viewing compounds NASA had set aside for them. The viewing spots were farther away from the launch pad than ever before, because an acid rainfall from the SRBs' exhaust had spread over a wider area than expected in April. The Kennedy Space

Center had even gone to the expense of providing $20 plastic automobile covers for employees' automobiles parked in the fallout area.

At T minus 9 minutes, the countdown clock stopped. There was a billboard-sized clock display in front of the press grandstand counting down in brightly lighted digits. A groan went up from the crowd.

Hugh Harris, the Kennedy Space Center public affairs commentator, explained that ground-based computers had stopped the count because of a drop in oxygen pressure in two of the three orbiter tanks supplying the fuel cell batteries. The two tanks had lost a pound from the nominal pressure of 800 pounds per square inch.

Realizing that there had been some interruption in topping off the tanks to replenish boil-off during the count, Firing Room controllers instructed the ground computers to ignore the low pressure signal and restarted the countdown clock. Cheers.

The lighted digits were blinking toward liftoff when the clock stopped again at T minus 31 seconds. At this point, two events occurred. *Columbia*'s computers took over control of the countdown, and the third liquid oxygen tank serving the fuel cells dropped in pressure. Having no order to ignore that event, the computers stopped the clock.

The Firing Room recycled the countdown to T minus 20 minutes, but the clock remained stopped as controllers tried to figure out how to bypass the ship's computers. On the flight deck, Engle and Truly waited in silence. Outwitting *Columbia*'s electronic brain was not easy.

Hugh Harris came on the public-address system again. The count would be picked up at 9:35 A.M., he said. That moment came and nothing happened. The hold had given the launch team time to examine another problem. Two of the ship's three auxiliary power units, which energize the vehicle's power steering system, were displaying excess lubricating oil pressure.

Page then acted decisively. He scrubbed the launch.

November 12

Inspection of the power units confirmed an early suspicion that the oil filters had become clogged with im-

purities. The filters were changed, and after exhaustive inspection of the units, the launch was rescheduled for 7:30 A.M. November 12.

Meantime, the oxygen pressure drop which had inaugurated the chain of delays leading to the scrub was traced to a faulty regulator in the ground support equipment. It was a space age version of the old refrain: For want of a nail, the battle was lost, or least postponed. The regulator had inhibited the ground support system from topping off the oxygen tanks.

Engle and Truly flew back to Kennedy from Houston November 10 and once more prepared to take *Columbia* into orbit. Next day, a new problem appeared. An electronic device in the orbiter's data processing system failed. It was one of 19 multiplexer-demultiplexer units, or MDMs, distributed throughout the vehicle. They transfer inputs from the ship's sensing systems to the computer-brains and output from the computers—commands—to the vehicle flight control systems.

Kennedy technicians replaced the defective MDM unit with another flown up from the Marshall Space Flight Center at Huntsville, Alabama, but the replacement unit didn't work, either.

Two units were then taken out of the second orbiter—*Challenger*—at Palmdale, California, and flown to Kennedy. They arrived at 9 P.M. November 11, and one of them was installed, connected, and checked out satisfactorily by midnight.

NASA reset the launch time once more and announced that *Columbia* would go November 12 at 10 A.M.

Once more, the crowds assembled on the beaches, causeways, and roadsides near the Space Center, busloads of VIPs rolled into the preferred viewing sites, a shrunken press corps gathered, and the countdown clock ran optimistically.

And once more the clock stopped at T minus 9 minutes.

But this time, the problem was purely psychological. George Page wanted the firing team to relax, look around, make sure everything was right. He wanted the Firing Room people to catch their breaths. One of the great moments of truth of the space age was only nine minutes away—the first reuse of a space vehicle. And the first

Firing Room 1 during the countdown of STS-2. Seated center is George F. Page, launch director. Immediately to his left stand Richard G. Smith, Kennedy Space Center director, and Tom Utsman, KSC technical support director.   (NASA)

John W. Young, flying the Shuttle training aircraft, took this picture of *Columbia*'s second launch November 12, 1981, with a hand held camera.   (NASA)

reuse of a rocket engine—the Space Shuttle Main Engines. The engines were the heart of the test.

Then it came. At 10 minutes past 10 A.M., the five engines that launch *Columbia*—the three main engines in the orbiter and the twin solid rocket boosters—ignited seconds apart with an Earth-shaking roar.

Again this remarkable vehicle rose smartly into a clear blue sky, its twin boosters and external tank giving it the profile of a medieval turreted castle spouting flame. The graceful ascent was visible for several minutes, and those with field glasses reported they could see the SRB's separate as *Columbia* and its propellant tank continued out over the Atlantic Ocean to become a dwindling, bright yellow spark.

It had been done. The second launch passed into history. Controllers at the Johnson Space Center in Houston took control of the mission. But . . . no sooner had Engle and Truly reached orbital altitude than they were alerted by an alarm.

The number 3 power unit (APU) was registering ab-

Two veteran flight directors monitor the STS-2 mission at Mission Control. Left is Eugene F. Kranz, flight operations director, and Christopher C. Kraft, Jr., director of Johnson Space Center.   (NASA)

normally high temperature, a forecast of trouble with the APUs. But the ascent had gone well. When the main rocket engines shut off at 8 minutes, 42 seconds into the flight and the fuel tank was jettisoned, *Columbia* was moving at 25,668 feet per second—just one foot per second slower than planned velocity.

Lubricating oil in APU 3 was overheating. The water cooling system had frozen. Truly turned off the APU, and shortly thereafter the system thawed out.

Another cooling problem appeared in APU 1. A cooling system designed to prevent vapor lock from forming in the fuel pump while the APU was turned off in orbital flight was not working.

This was not a critical failure, because the APU might work anyway when it was turned on for the return to Earth. In any event, two APUs could provide sufficient hydraulic power to run the ship's power steering system during descent.

The real problem was fuel cell battery 1. The first sign that it was not operating properly was an abnormal pH reading, indicating rising acidity in one of the battery's 64 fuel cells. This condition was reported by the crew as *Columbia* passed over Australia on its second orbit. It was an indication that the flow of hydrogen into the cell was being blocked. The hydrogen combines with oxygen in the presence of a catalyst, potassium hydroxide, to yield direct electrical current, with pure water as a by-product. The water is stored for the crew.

Telemetry signals from the spaceship, relayed to Houston from the Australian tracking station at Yarragadee, showed that voltage was dropping in fuel cell battery 1. At Houston, the capsule communicator, Dan Brandenstein, instructed the crew to perform a routine purge of fuel cell batteries 2 and 3, but to do nothing to number 1. The batteries are flushed out periodically to prevent a buildup of contaminants that could block the flow of hydrogen and/or oxygen into the cells.

The crew, meanwhile, had fired the orbital maneuvering system (OMS) engines twice to establish *Columbia* in a 120-nautical-mile orbit. Two more burns were required to lift the spacecraft into its 137-nautical-mile (150 statute miles) planned cruising altitude.

Before the burns were executed, however, controllers were polled by the early shift flight director, Neil Hutch-

inson, on the status of all spaceship systems.* With two of the three fuel cell batteries working, there was sufficient electrical power to operate the ship, the experiments in the payload bay, and the arm, but the rules of the mission required that if one battery failed, the flight would be reduced to 54 hours.

Three hours and 19 minutes into the flight, Hutchinson authorized a "Go" for the crew to stay in orbit, but Brandenstein repeated the order not to purge fuel cell battery 1.

Engle and Truly struggled out of their space suits and settled down to what they hoped would be the full five-day mission they had been training for.

But as the full orbital altitude was reached with additional OMS burns, an assessment of the fuel cell problem showed that *Columbia*'s electric power system was irrevocably crippled for this mission.

The obstruction of hydrogen flow into just one of battery 1's 64 cells was allowing water to accumulate. If it spilled into the battery's manifold, it would be exposed to a voltage that would reverse the power reaction and break down the water into hydrogen and oxygen. The electrolysis would allow an explosive mixture of hydrogen and oxygen to accumulate and one spark could ignite it to blow up the battery and damage the ship. This was the "worst case" prospect.

Also, there was danger that the potassium hydroxide catalyst would get into the overflow water, and the water would contaminate the crew's drinking supply.

Brandenstein passed down instructions to the crew, as *Columbia* came within range of the tracking station on Guam, to close the hydrogen and oxygen valves of the battery and then restart it, leaving the valves closed. This would use up the gases remaining in the cells.

"That way we get the hydrogen and oxygen out and safe [sic] the fuel cell good," Brandenstein said.

A short time later, he gave the crew the bad news.

"We're still looking down the road with this fuel cell being safed," he said, "and not being able to start it up. We will probably be looking at the minimum mission."

That was the 36-orbit, 54-hour flight that Young and Crippen had flown in April.

---

*Twenty-four-hour flight control at Houston was divided into three eight-hour shifts.

The STS-2 crew took this picture of *Columbia*'s cargo bay, with OSTA-1 experiments in the foreground.   (NASA)

The final OMS burns were made at 7 hours, 45 minutes and 8 hours, 33 minutes, using first the right engine and then the left engine only. They put *Columbia* in an orbit slightly higher than planned, 149 by 139 nautical (161 by 158.9 statute) miles, at an inclination of 38° to the equator.

At 10 hours mission elapsed time, Hutchinson described a compressed mission plan that had been prepared in case the flight had to be abbreviated. It would allow most of the flight objectives to be met by speeding up crew activity and working more hours.

The OSTA experiments were self-operating, except for the lightning photographs, which had to be taken by Truly or Engle. With its payload doors open, *Columbia* had been flying upside down for hours with respect to the ground, so that the OSTA scientific instruments in the payload bay and the crew were facing the Earth.

The Minimum Mission

Even though the crew had been warned that the minimum mission would be flown, there was no certainty about it at Houston. In fact, Hutchinson suggested at the evening briefing of November 12 that the flight might go longer than the 54-hour limit, despite mission rules.

The minimum mission plan, Hutchinson said, "does not mean necessarily that we're going to come down tomorrow or the next day or the next day. What it does is put us in a posture that if we decided tomorrow or the next day or the next that we want to shorten the flight, we will have already put behind us the most important things that we had to do."

Was it still possible to run the five-day mission? asked Morton Dean of CBS.

"Yes, sir," replied Hutchinson. "We're going to see where we are tomorrow afternoon and decide."

Thunderstorms were reported over South Africa, but the crew was not able to get lightning pictures as *Columbia* flew over. Good news for a change came on the morning of November 13. All three APUs were pronounced fit for entry.

An extra tidbit of public information was disclosed overnight. *Columbia*'s weight was calculated at 221,983 pounds. It still had most of its fuel and water on board.

Another thunderstorm was reported over Australia. Bernard Vonnegut of the State University of New York, Albany, the scientist directing the lightning flash experiment, had said that there is ten times more lightning observed by satellites over land than over water. Scientists didn't know why. Also, there is lightning on other planets that somehow produce it without rain.

There are many questions about lightning, but not many answers, Vonnegut said. Maybe Engle and Truly could illuminate the mystery by photographing it from above the clouds with their 16-millimeter camera at 24 frames a second.

Over Australia, Truly deployed the RMS arm and moved it about. It worked perfectly, he reported. The grasping hand was not used lest it get stuck in the payload bay, in which case a crewmember would have put on a space suit and go outside to free it. The test flight program was not yet prepared for that.

As *Columbia* passed over Florida on orbit 17, a woman's voice was heard calling the spaceship from Houston. It was a first. The voice belonged to Sally Ride, a member of the first contingent of women astronauts and the first woman to serve as capsule communicator.

Destined to be America's first woman in space, astronaut Sally K. Ride, a mission specialist, talks to STS-2 commander Joe Engle through the spacecraft communicator panel at Mission Control. (NASA)

Using the 70mm Hasselblad camera, Truly took this picture of the Canadian-built remote manipulator arm flexed at the elbow to pick up a package in the aft cargo bay with its end effector, a mechanical hand. A television camera is mounted near the elbow on the forepart of the arm. (NASA)

She sounded crisp and friendly. When Truly made a brief report, she acknowledged; "Okay, sounds good."

"You sound mighty good, too," Truly replied.

"Okay," said Ride, "Go ahead."

"The arm is out and apparently getting some good TV through the wrist and arm cameras," Truly reported. "The elbow camera scanned *Columbia*'s exterior, and also there was a shot of the arm deployed and of the Earth from the camera on a forward bulkhead of the payload bay. So far, the arm has been really operating smooth. In general, it's [*Columbia*] a remarkable flying machine and it's doing exactly as we'd hoped and expected."

"Hey! That's great news, Richard," Ride said.

At 11 A.M., NASA announced its decision to restrict the mission to 54 hours and to land at Edwards Air Force Base at 1:22 P.M. pacific standard time, November 14.

Ride called up the news to the crew, and they took it philosophically. The arm was folded back into the payload bay to allow the radar camera freedom to resume photographing the surface.

All in all, the camera ran for eight hours during the flight and photographed about 3.8 million square miles of the United States, Europe, Africa, Australia, South America, and Indonesia. Dr. Charles Elachi of NASA's

Mission Control Operations Room (MCOR) at the Johnson Space Center, Houston, during the flight of STS-2. The tracking map (left) shows the position of *Columbia* over the Hawaiian Islands. An onboard television view of the open cargo bay is being transmitted to Houston via Hawaii tracking on the center screen. (NASA)

Seated with telephone at Mission Control, President Reagan calls the crew of STS-2 and jokingly asks if *Columbia* could pick him up in Washington and give him a lift to California. Amused are capsule communicator Daniel C. Brandenstein (seated left) and, standing left to right, capsule communicator Terry Hart, Deputy NASA Administrator Hans Mark, NASA Administrator James M. Beggs, and Kraft.   (NASA)

Jet Propulsion Laboratory, the investigator, later reported that the camera photographed a 45-minute swath 10,000 miles long, from Spain through the Mediterranean Sea to India, Indonesia, and Australia. It was the longest continuous stretch of imaging data NASA has acquired from a space-borne instrument.

In the evening of the second day, the crew was hailed by a visiting "capsule communicator"—President Reagan, who dropped in to Mission Control to see how things were going.

Aside from the dead fuel cell battery, *Columbia* was behaving. Or it was for a time. Presently, Engle reported that the cathode-ray tube display in front of his chair had failed. Houston said he could use one of the other three still working.

Since that was inconvenient, Engle picked up a screwdriver and replaced the tube with a live one from the aft flight deck. This orbital bulb snatching marked the first Shuttle repair in space.

"It's working," Engle reported proudly.

"All right," said Houston. "You get the golden wrench award."

"And if you want us to fix that fuel cell," Engle added, "we can run on out to five days."

Autoland

The second night, *Columbia* was cruising around the Earth at a velocity of 25,436 feet per second in a slightly oval orbit, 167.8 to 160.5 miles high. The altitude had been changed by tests of the space engines. Both crewmen had reported that the bow thrusters shook the ship when they fired.

On orbit 25, Houston called to advise the crew that the payload pallet should be secured early the next day. The instruments were turned off in the morning. Engle and Truly, who slept both nights in their chairs, disdaining a sleeping bag that had been rigged for testing, were roused early November 14 with song—a rendition by the Muppets of "Columbia, the Gem of the Ocean." You never hear Beethoven, Bach, or Mozart in space.

The weather forecast for Edwards, high in the Mojave Desert, offered some broken clouds at 25,000 feet, no rain, and winds within safe limits. Houston advised the crew to turn up the APU heaters to warm the lubricating oil. The flight recorder was set on continuous record.

Mission Control viewed the data from the recorder as critical now, because the recorder had failed during the first mission and information on temperatures and pressures had been lost during the descent across the Pacific while the radio was blacked out.

The need for these data became a factor in curtailing this mission, according to Christopher C. Kraft, Johnson Center director. Kraft had emphasized that the control team didn't want to chance another system failure that would cause the loss of descent data.

The payload doors were closed over Australia. On the 36th orbit, Engle turned the ship around as it passed the tracking station on Ascension Island in the South Atlantic. Now the OMS engines were pointed in the direction of flight.

They were fired for 2 minutes, 55 seconds as *Colum-*

Engle exercises on the treadmill to preserve muscle tone, which otherwise tends to deteriorate in microgravity.   (NASA)

Truly studies a bundle of floating teleprinter copy.   (NASA)

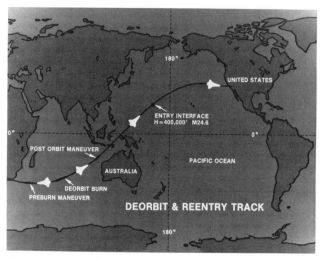

The reentry and descent of *Columbia* are charted. The crew turns the ship around so that its orbital maneuvering engines point in the direction of flight. When they fire, the ship's velocity is reduced. The deorbit burn over the Indian Ocean put the ship in a descent orbit on which it glided across the Pacific Ocean to land in California.   (NASA)

bia, flying tail first, passed over the Indian Ocean. The retrothrust cut velocity by 313.4 feet a second, changing the orbit to an ellipse with a high point (apogee) of 141 (nautical) miles and a low point (perigee) of minus 2 miles.

Once again, the hills and roads around Rogers Dry Lake became black with crowds and traffic, as more than 200,000 people converged on Edwards Air Force Base to watch *Columbia* land.

John Young went up in a NASA Gulfstream trainer to look at the weather, as he had done at Kennedy before the launch. He recommended that, because the winds were strong, Engle make the sweeping turn at the point called the heading alignment circle (HAC) manually, rather than allow the autoland system (autopilot) to control the turn.

The HAC is the area over California where the orbiter swings around to align itself with the runway on the dry lake. With stick control, Engle could reduce wind and gravitational stress on the spaceship.

The flight plan called for landing on runway 15, which runs north-south, so that *Columbia*'s performance in a crosswind landing could be tested. However, John Young warned that the winds would be stronger than anticipated. Mission Control switched the landing to runway 23, lying at right angles to runway 15. *Columbia* would thus land into the wind.

Crossing the equator on the 37th orbit, *Columbia* hit the sensible atmosphere at 413,719 feet and entered the usual radio blackout as it zoomed across the Pacific toward California. Just before the ship emerged from blackout, radar picked it up at 653 miles off the coast, descending through 188,000 feet.

The ship was moving at a velocity of Mach 15 but was 25 miles south of the planned course. The guidance error was quickly corrected automatically as blackout ended and radio guidance was resumed.

A voice came through static from *Columbia*. "Houston, *Columbia*. Air data look good on board." Conversation between the spacecraft (SC) and the capsule communication (CC) continued in rapid fire.

CC: 82,000 feet. Mach 2.5. 57 miles. Everything is looking right on the money, there, Joe, and we have a wind update for you and a weather update. You've got a very thin layer at 25,000. The winds aloft are as briefed on the ground, 18 knots, gusting to 24. You got 60 miles vis [visibility] underneath. Over.

SC: Very good. Sounds like a good old ready day.

CC: Yes, sir. Out of 68,000 [feet]. 39 miles range, Mach 1.5. 50,000 feet, Mach 1. Range 27 miles. You're tracking right down the line.

The crowds at Edwards heard a sharp double sonic boom. *Columbia* was coming in.

At the heading alignment circle, Engle took manual control of the ship and made a sharp turn, dropping slightly below the intended glide path. He lowered the body flap to reduce speed.

CC: 280 knots at 18,000 feet.

SC: Okay, speed brakes starting now.

CC: Roger. Still just slightly low on energy. Looking okay. 9 miles. 13,000.

Engle closed the speed brakes, aware that the ship's energy (velocity) was a bit low. This reduced air resistance.

CC: Roger. Slightly below glide slope. You're below glide slope. You have a Go for autoland.

Engle relinquished manual control to the autoland system, now in contact with a microwave landing beam on runway 23. The beam would guide the autopilot to a precise landing.

As soon as autoland took control, it pitched up the spaceship to the planned glide slope. The fly-up, though brief, had the effect of reducing velocity, causing *Columbia* to land short of the precise touchdown point marked for it.

CC: We're about a minute away from touchdown. 3,500 feet. 250 knots.

SC: Okay, 2,500 feet. Speed brakes closed.

With the brakes off, *Columbia*'s descent speed hit 270 knots. A chase plane appeared off the port wing to escort it down. Test Manager Donald K. (Deke) Slayton was at the controls. He called off the the final figures of the descent.

"Columbia clear to land on lake bed 23. 100 feet, 50, 30, 20, 10, 5, 3. Touchdown! Nosegear 15, 10, 5, 3. Touchdown! Welcome home!"

SC: Thanks, Chase.

Engle (left) and Truly walk away from a successful test flight as they give a quick resume of the 54-hour mission to Flight Operations Director George W. S. Abbey.   (NASA)

The Radar Time Machine

Months later, it was disclosed that the Shuttle Imaging Radar–A (SIR-A) had scored a remarkable breakthrough in the technology of remote sensing. As *Columbia* passed over North Africa on its 27th orbit on November 14, 1981, SIR-A took pictures of the Eastern Sahara, a desert as dry as the deserts of Mars. When the images were processed, they revealed not the sea of sand that exists today but a vast network of ancient river valleys and streambeds formed millions of years ago. Radar had penetrated one to two meters (1.2 to 2.2 yards) of dry sand and mirrored the prehistoric landscape that lay beneath it.

Although the power of radar to penetrate dry sand was predicted by theory, the actuality came as a stunning surprise to the SIR-A experimenters. After the film was processed at the Jet Propulsion Laboratory in Pasadena, it was sent to the Astrogeologic Studies Branch of the U.S. Geological Survey (USGS) at Flagstaff, Arizona, for interpretation.

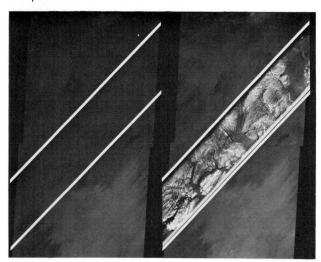

This photo montage contrasts a Landsat picture of the Egyptian Sahara (left panel) with the STS-2 SIR-A image of the same part of the desert stripped diagonally across the right panel. The strip, 30 miles wide, reveals a prehistoric drainage system, which millions of years ago consisted of rivers and streams in what is now the driest region on Earth. The Shuttle Imaging Radar–A (SIR-A) "camera" was able to photograph the ancient terrain below three to six feet of sand.   (NASA-JPL)

The SIR-A images of Egypt and the Sudan were first examined by Carol S. Breed, a geologist who had made field studies in the region. At once, Dr. Breed realized that she was seeing in the images a buried landscape that had been part of a great watershed in a bygone epoch of geologic time.

The Sahara's hyper-arid core is called the Arbain desert. SIR-A investigators said it was the driest region on Earth. The absence of moisture allowed radar to penetrate the sand and be reflected by the denser material below to the SIR-A antenna to form the image.

Images of the Arbain acquired by the Earth resources satellite *Landsat 2*, equipped with multispectral scanner and vidicon imaging devices, showed nothing but a sea of sand. Dr. Breed and colleagues had previously studied networks of ancient stream valleys emanating from a high plateau near the Libyan border. On Landsat images, these valleys disappeared beneath the sands of the Arbain. Now SIR-A had found them.

The next step was to confirm the SIR-A results on the ground, said Gerald G. Schaber of USGS-Flagstaff. USGS and JPL scientists organized an expedition to the Egyptian desert in September 1982. They were joined by scientists of the Egyptian Geological Survey, who brought along a pick-and-shovel crew to dig pits and search for river sediments that would verify the radar scenes.

Before the workmen had dug a meter's depth they turned up river gravel. Mixed with it they found stone tools. The scientific team cited archeological evidence of human occupation dating back 200,000 years or more. In a report published in *Science*, the journal of the American Association for the Advancement of Science, the scientific team said that the buried valleys were relics of integrated river systems that had drained the eastern Sahara in the Tertiary period (beginning about 40 million years ago).[2] With the onset of the Quaternary 2 million years ago, the climate changed and the region became progressively drier. Intermittent years of rainfall alternated with times of aridity. In one time, the Sahara had been a grassy plain crisscrossed by streams and dotted with lakes and ponds; in another, it became a dry expanse of rock and sand.

The present hyper-arid climate set in about 5,000 years

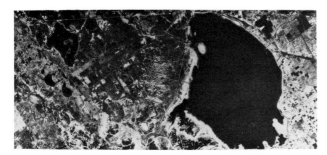

SIR-A images Lake Okeechobee and smaller lakes in south Florida. (NASA-JPL)

The SIR-A image of the Los Angeles basin. Smog, clouds, or darkness are no deterrent to radar imaging. The microwave energy penetrates smog and clouds and creates its own illumination. (NASA-JPL)

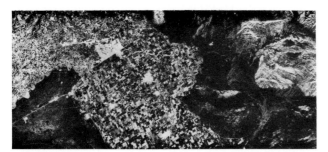

SIR-A shows a 60-by-30-mile swath of Southern California's Imperial Valley. The checkered pattern is formed by cultivated fields. (NASA-JPL)

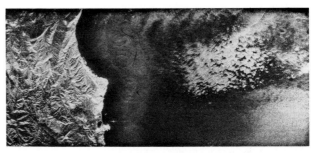

Despite cloud cover and darkness, this image of a 75-by-30 mile area of the Mediterranean Sea and the eastern coast of Sardinia (left) was acquired by SIR-A. Bright, mottled areas in the sea (right) represent choppiness. (NASA-JPL)

SIR-A captures the California coast from Point Concepcion (far left) to Ventura (right). (NASA-JPL)

ago. People gradually abandoned the region. Now the driest part of the Sahara, the Arbain, occupies about 77,000 square miles in southwest Egypt and northwest Sudan.

Before the onset of hyper-aridity, the report said, the region was inhabited by early man in the Lower Paleolithic, by Neanderthal man in the mid-Paleolithic, and by successive Neolithic cultures of modern man.

The report of the SIR-A investigators is direct and unqualified. It said: "The SIR-A radar saw below the sand sheet a dramatically different and predominantly fluvial subsurface terrain known previously only to stone age people."

The penetration of SIR-A is limited to arid regions, the experimenters explained. Surface moisture attenuates the signal. Other deserts where SIR-A imagery was acquired were the Taklamakan, Badain, Jaran, Ulan Buh, and Mu Us in China; the Kara Kum in the Soviet Union; the Rajasthan in India; the Thar in Pakistan; desert areas of central Australia; and the An Nafud in Saudi Arabia. An image of the Saudi An Nafud published by JPL looked remarkably like those of the subsurface Arbain, but the nature of the An Nafud lineaments was not identified.

SIR-A images were obtained in strips, each representing a swath 31 miles wide on the surface. The images covered 3.8 million square miles along *Columbia*'s ground track, which passed through all the continents except Antarctica.

Implications of SIR-A imaging for extraterrestrial exploration were cited in the *Science* report. It observed that the ancient fluvial landscape under the Arbain sands "resembles features of probable fluvial origin in the northern plains and equatorial regions of Mars."

"We can only speculate," the report added, "on the potential of imaging radar to reveal subjacent topography through the dry or frozen eolian veneer that mantles many areas of Mars."

The SIR-A experiment was conducted by Charles Elachi, principal investigator, JPL, and coinvestigators Walter E. Brown and R. Stephen Saunders, JPL; Louis Dellwig, University of Kansas; Anthony W. England, NASA/Johnson Space Center; Max Guy, Centre National d'Études Spatiales, France; Harold MacDonald, University of Arkansas; and Dr. Schaber, USGS. Dr. Breed was guest investigator.

Eight Days in March

Columbia's third orbital test flight (STS-3) was projected for 7 days, 5 hours. It was designed to determine the spaceship's response to extremes of heat and cold in space; how well its electrical, life-support, and propulsion systems would function on a mission lasting 173 hours; and the utility of the orbiter as a scientific observatory in its own right, without the addition of Spacelab. A British Aerospace pallet containing an array of scientific experiments was emplaced in the cargo bay.

The flight plan called for the use of the Canadian-built remote manipulator system (RMS) to pick up two experiments from the pallet. The 50-foot arm would lift them 45 feet above the cargo bay to monitor the space environment. The arm would then return the instruments to the pallet.

One of the experiments was a plasma detector, a cylinder 42 inches long and 26 inches in diameter. It had a mass of 350 pounds on Earth, but in orbit, of course, it was nearly weightless.

In space physics, *plasma* is the term used to describe an ionized "gas" consisting of electrically charged particles—ions and electrons. A plasma region called the ionosphere exists above the sensible atmosphere at Shuttle flight altitudes. It is a region where electrons are stripped from atoms of oxygen, nitrogen, or hydrogen by sunlight, and at certain densities it forms invisible mirrors that reflect AM radio waves back to the Earth's surface.

At an altitude of 150 miles, *Columbia* would be cruising through the ionosphere. The plasma experiment would measure its density, and other detectors would diagnose the electrical effects of the ship's motion through this sea of electrically charged particles.

A second experiment the arm was to lift above the vehicle was designed to monitor gases and water vapor emitted by the spaceship itself—contaminants induced by *Columbia* into the space environment near the Earth. This was a desk-size package called, appropriately enough, the Induced Environmental Contamination Monitor.

Preparing *Columbia* for the third flight took ten days

less than processing it for the second had, a gain on the learning curve of reusing the Shuttle. Early in February 1982, the Kennedy Space Center launch directorate set March 22 as the target launch day. This time the orbiter was launched on the day scheduled. But it was an hour late.

The 73-hour countdown had been running like a watch until seven hours before the 10 A.M. EST launch time. Before dawn, a heater in the ground servicing system failed to turn on and heat gaseous nitrogen. The gas was pumped to purge fuel and oxidizer lines from the propellant tank to the main engines. It took an hour for a technician to find and repair a faulty switch. The launch was rescheduled to 11 A.M.

This gave the crew an extra hour of sleep. Commanding the mission was Marine Corps Colonel Jack R. Lousma, 46, an astronaut since 1966. He had been pilot of the 59½-day mission of *Skylab 3* in 1973. The Shuttle pilot this trip was Air Force Colonel Charles Gordon (Gordo) Fullerton, 45. He had been a member of the test crews that flew *Enterprise* on glide tests in 1977.

Northrup Strip

During the week before the launch, uncertainty developed among the Shuttle directors as to where *Columbia* would land. Persistent rains had thoroughly soaked the lake-bed runways at Edwards Air Force Base. Directors were considering the Northrup Strip, an experimental aircraft runway system at the White Sands Missile Range in New Mexico, as an alternative. Shortly before the terminal countdown started, the directorate received a forecast that the lake bed in California would continue to be water soaked for at least the week of the flight. The Johnson Space Center announced that *Columbia* would land at White Sands.

Northrup Strip had been marked out for years as a landing site on the gypsum flats in the south central region of the 4,000-square-mile missile range. The main runway, a strip 300 feet wide and 35,000 feet long, plus a 3-mile overrun, extended roughly north and south (350° and 170°). Because of prevailing winds, the north-south

runway—17—was most frequently used. It was intersected by a strip running roughly northeast-southwest (50°–230°), Runway 23.

Ground equipment for a Microwave Scanning Beam Landing System had been installed on runway 17, but not on 23. Other equipment there included a TACAN (tactical air navigation beacon), an S-band communications system in a van, and a derrick to hoist *Columbia* atop its Boeing 747 carrier aircraft for the return to Florida.

The decision to land in New Mexico instead of holding the flight at Kennedy until Edwards dried had interesting ramifications. It betrayed the anxiety of the Shuttle directorate to bring the vehicle into operational status on schedule, that is, by fall 1982.

The new administrator, James M. Beggs, and his associate administrator for Space Transportation Systems, Major General James A. Abrahamson, seemed to be determined to complete the flight testing without delay. An aspect of their concern was the growing competitive threat of the European Space Agency's Ariane launcher. Its managers were underbidding NASA for communications satellite payloads.

The New Mexico decision also reflected the caution of a new NASA administration. It was unwilling to risk a landing on a concrete runway either at Edwards or at Kennedy without a prior test of the vehicle's landing performance in a crosswind. Flight directors thus were committed to test a crosswind landing on a lake-bed runway at Edwards or the alkali flats at White Sands before committing the orbiter to a three-mile strip of concrete.

Although White Sands had been considered only as an alternate site, to be used in an emergency, the Northrup Strip had been fairly well equipped. The autoland system could be tested on runway 17 if weather permitted *Columbia* to land on it.

On the morning of March 19, a 23-car train loaded with service equipment for *Columbia* departed Edwards for Holloman Air Force Base, New Mexico. Trucks were lined up at Holloman to haul the equipment 23 miles to Northrup. The shipment included mobile carriers equipped to provide ground cooling and to purge the ammonia coolant system on board and the space engine fuel lines. A second train of 15 cars left Edwards the next day.

A temporary complex of nine service, storage, and office buildings and an aircraft hangar was set up near the strip. Communication lines were provided for news media correspondents.

By launch day, 550 NASA and private contractor people had been transported to White Sands from Florida and California. It was a massive operation, accomplished with some efficiency. Although NASA officials pleaded inability to reckon the excess cost of this operation, they insisted it was less costly than allowing *Columbia* to sit on the launch pad three or four weeks waiting for Edwards to dry out.

See You in Madrid

The terminal countdown began shortly after sunrise. Lousma and Fullerton boarded *Columbia* at 8:55 A.M., checked their communications and the guidance and navigation system and set the switches for launch. The hatch was closed. The cabin was pressurized and scanned for leaks.

At T minus nine minutes, the ground launch sequencer took over the count and began to measure a thousand launch criteria. The orbiter access ramp was retracted. The crew was committed to the ship. The propellant tank was pressurized and the valves closed.

T minus 16 seconds. The sound suppression water system was turned on, spraying 70,000 gallons a minute into the flame pits to reduce the roar and exhaust backwash. T minus 6.8 seconds. *Columbia*'s main engines were ignited in a sunburst of yellow flame and white clouds of steam. As the engines roared up to a million pounds of thrust, the giant solid rocket boosters erupted with billows of smoke. *Columbia* rose on 6.9 million pounds of thrust and twin pillars of fire from the boosters. Jake Miller, the Brevard County sheriff, estimated that three quarters of a million people had gathered on the roads, riverbanks, beaches, and causeways around the Space Center to watch. *Columbia* rose steeply and nearly

vertically, vanishing behind distant cirrus. Then it pitched over and rolled so that the crew was turned to a heads-down position aligned with the launch azimuth of 81.36° as measured clockwise from north. *Columbia* was then headed toward a 150-mile orbit at an inclination of 38° to the equator. Once *Columbia* had cleared the tower, launch control shifted to Houston.

Voices filtered over the public-address system.

SC (spacecraft): Oh, what blue skies.

PAO (public affairs officer): Twenty seconds. Thrust looks good. Roll maneuver completed.

SC: (Lousma) The first part of this ride is a real barn burner.

PAO: Two minutes, 45 seconds. *Columbia* has two-engine reach to Rota [This meant that if one engine failed, the ship could reach the U.S. Naval Air Station at Rota, Spain, on two engines].

SC: We got some flakes going by the window. We got some right after liftoff. Blue out here. Lots of light, little flakes.

Observers paid little attention to that report then. It was not until the next day that they realized the little white flakes were pieces of the heat shield tiles breaking off.

The crew was obviously enjoying the ride.

On the ground, meantime, launch director George Page reported that all aspects of the launch had been nominal. Chided about launching an hour late, he looked on the bright side. "For the first time, we did it on the same day we planned to," he said.

The solid rocket boosters burned out at 2 minutes, 2 seconds and parachuted into the Atlantic, where the United Technology Corporation's recovery ships, *Liberty* and *Freedom*, waited to pick them up and tow them back to Port Canaveral for refurbishment and eventual reuse.

The million-dollar nozzle plug, described in chapter 9, had been abandoned in November 1981. It was the self-propelled, guided "missile" that when launched from the recovery ships, swam to a booster casing and, under control from the ship, plugged the nozzle throat. The device was designed to pump seawater out of the 150-foot rocket casings so they could float horizontally for

The solid rocket booster casing bobs vertically in the Atlantic Ocean 160 miles east of Cape Canaveral as the *Liberty* prepares to retrieve it. The parachutes have already been recovered, and one of the divers facing the empty rocket is preparing to install the plug.   (NASA)

Liberty's sister ship, *Freedom*, has picked up the second SRB. (United Space Boosters)

Liberty's retrieval crew makes fast the tow line to the floating SRB casing.   (NASA)

The retrieval ship has the booster casing in tow and heads back to Port Canaveral, Florida.   (United Space Boosters)

towing. Difficulties with the nozzle plug during booster recovery on STS-1 persuaded the recovery contractor to use divers to emplace the plug. A gallant engineering development was shelved, at least for the time being.

Seven minutes into the flight, *Columbia* pitched over and went into a shallow dive. The maneuver, peculiar to a winged vehicle, increased its velocity even though it decreased altitude. Then *Columbia* resumed its climb out of the atmosphere.

Lousma began reporting some overheating in auxiliary power unit 3. Shuttle Control had been watching it on telemetry. The other two APUs were running smoothly, providing power for the hydraulic actuators that moved the aerosurfaces and swivelled the main engines for steering during the ascent to orbit.

Here they are, close up, with their big "fish." *Freedom* is in the foreground and *Liberty* off to the right.

(United Space Boosters)

The booster disassembly facility at Hangar AF, Cape Canaveral Air Force Station. The retrieval ships are docked and the boosters are moving through the disassembly and refurbishment process to be used again on a later flight.   (NASA)

Back in port, *Liberty* tows the booster casing through the channel to Kennedy Space Center's refurbishment facility.

(United Space Boosters)

At 7 minutes, 38 seconds, Terry Hart, the capsule communicator (CC) at Houston, called:
"We recommend you secure APU 3."

SC: Okay. Shut down APU 3.

Although the APUs were due for shutdown on orbit, the loss of one posed a potentially serious problem. Mission rules required that at least two APUs be working, or in working order, to continue the flight. If another one failed, the flight would be terminated.

Controllers suspected that the cause of overheating was a freeze of the lubrication system. A power unit had failed on the second flight for the same reason, but had thawed out and functioned nominally on the descent. APU 3 might again be in working order when the lubrication system thawed. Mission Control advised the crew of this possibility. Why it had frozen during ascent was a mystery.

The main engines cut off at 8 minutes, 30 seconds, and there was no further need for the APUs until entry, 7 days ahead. Lousma throttled the engines back to 68 percent of their rated thrust a few seconds before cutoff.

They had worked beautifully, the most powerful hydrogen-oxygen engines ever built. Now *Columbia* was coasting toward orbit. The big propellant tank was jettisoned by explosive bolts. It tumbled away in a long arc, to splash into the Indian Ocean. *Columbia* was 635 nautical miles downrange.

At Houston, the on-shift flight director, Tom Holloway, sent a "Go" for the first burn of the twin orbital maneuvering engines, to boost the ship toward its final orbital altitude. The crew maneuvered the ship to the prescribed burn attitude and fired engines at 10 minutes, 43 seconds elapsed time.

"Burn looking real good," Lousma reported. "It was on time. We'll see you in Madrid."

Columbia flew beyond the range of the Bermuda tracking and relay station. Communications would be resumed at Madrid, just a few minutes away.

Firing the OMS engines for 1 minute, 27 seconds lifted *Columbia* into an oval orbit 130 nautical miles high at apogee and 46 nautical miles at perigee.

The crew came on the air again through Madrid: "We're in a 131 by 46 as you can see."

Telemetry gave the crew's heart rates: Lousma, 132 beats a minute during launch, and Fullerton, 92.

As *Columbia* reached the apogee of this initial orbit, the crew fired the OMS engines again to circularize the orbit at 130 nautical (150 statute) miles. The computer was switched to "Ops-2," the on-orbit flight program.

The Experiments

In addition to using the arm, the flight plan called for measuring the effect of heat and cold on the orbiter's structure. Outside the atmosphere, a surface turned toward the Sun becomes very hot, one away from the Sun ice cold.

On long flights to the Moon, the Apollo command and service modules were rotated slowly along their long axis like a hot dog on a spit—the barbecue mode, so that solar heating would be distributed. This method of controlling spacecraft temperature is called passive thermal control (PTC).

The test program called for the crew to fly *Columbia* tail toward the Sun for 30 hours, payload bay toward the Sun for 26 hours, and nose toward the Sun for 80 hours. For the balance of the flight, the vehicle would be rotated in PTC.

Thermal control has always been a problem in manned spacecraft. In *Columbia*, the thermal control system consisted of two freon loops and liquid heat exchangers and three subsystems for dumping the heat. The three subsystems were radiators attached to the inside of the payload bay doors, flash evaporation using boiling water, and an ammonia boiler.

The payload bay door radiators discharged heat into space in orbit, when the doors were open. If the heat loads become too much for them to handle, the flash evaporators cut in. The system was balanced to keep the cabin between 75° and 85° F. During ascent to orbit, the flash evaporators were set to go on 2 minutes, 25 seconds after liftoff. They would continue to discharge heat until the payload bay doors were opened and the radiators deployed.

The flash evaporators would be turned on again to reject heat after the doors were closed for entry and descent. When the orbiter reached an altitude below 120,000 feet, the boiling-water evaparation system was no longer efficient. Heat rejection from there on down was provided by evaporating ammonia through a boiler until the wheels stopped and *Columbia* could be connected to ground heat rejection equipment. Ammonia fumes gave the orbiter a commercial laundry odor on landing and required service crews to wear masks and protective clothing.

There were nine scientific experiments aboard STS-3, managed by the Office of Space Science and the Goddard Space Flight Center. Six were prepared by experimenters at five American universities and one by an investigator from a British institution. Goddard had two experiments and the Office of Naval Research one.

The Plasma Diagnostics Package (PDP) was designed by the University of Iowa and the contamination monitor by Goddard, with Air Force funding. The University of Florida had aboard an instrument that measured the plume produced by the emission of gases from the orbiter. Columbia University had an x-ray polarimeter to observe x rays in solar flares. The Naval Research Laboratory's experiment was an ultraviolet radiation monitor to observe ultraviolet radation from the Sun at wavelengths that do not penetrate the atmosphere. The University of Utah had a device to measure the electrical charge on *Columbia* as it passed through the ionosphere. The Vehicle Charging/Potential Experiment included an electron beam generator. The experiment would also measure changes in the ship's electrical characteristics when the beam was turned on. Goddard put aboard a large cannister with an elaborate temperature control system to protect temperature-sensitive instruments. It was being tested for future use. The British experiment was a micrometeoroid detector developed by the University of Kent, Canterbury.

In the cabin were two biological experiments. One, prepared by the University of Houston, was a miniature arboretum. It contained 96 small plants—slash pine, mung beans, and oats—in sealed containers. Its purpose was to determine the effect of low gravity on plant growth. The second experiment, designed by a Minnesota high school student in a NASA contest, attracted more media attention than any other. It consisted of observing the

flight behavior in low gravity of two species of insects, the velvet bean caterpillar moth and the honeybee drone. The insects were confined in glass cages in the crew compartment, and during the flight Fullerton photographed their activity. The moths fluttered about happily, but the bees huddled apathetically against the glass and eventually died.

Early in the flight, the crew reported the presence of an insect hitchhiker, which Lousma thought was a Florida fruit fly. It seemed to be buzzing around the flight deck for a while, then disappeared.

Two space processing experiments were carried in the crew compartment. One was a device for separating biological cells from a fluid by electrophoresis. The other was an automated reactor that manufactured tiny latex beads.

Electrophoresis appears to work more efficiently in low gravity than on the ground in separating cells and other biological material from a fluid matrix. The experiment was sponsored by Johnson & Johnson, the pharmaceutical house, and McDonnell-Douglas Astronautics. It is believed to have the potential of reducing the cost of extracting enzymes from cells, such as urokinase, used in medicine to dissolve blood clots from kidney cells.

The latex beads have medical diagnostic and treatment uses in diseases of the eye and in cancer. They can be used to carry drugs and radioactive isotopes to malignant tumors through the circulatory system. On the ground, beads can be made only up to three microns in size in identical lots. The reactor was able to turn them out in identical lots of up to nine microns in low gravity.

Jack R. Lousma, commander of STS-3, shows off passengers aboard *Columbia* on its third orbital flight. They are moths, bees, and flies caged in the box, subjects of a behavioral study of insects in free fall.   (NASA)

Tiles Off

Forty-one minutes after liftoff, the crew opened the payload bay doors to expose the radiators to space. Lousma began showing signs of motion sickness, which he had suffered during the early days of his 1973 flight on *Skylab*. Fullerton, who had taken anti–motion sickness pills, was thirsty. As *Columbia* passed over Australia, the commentary from Houston indicated that he was drinking his third eight-ounce bag of water. But he had no appetite, and neither crewman ate much food the first day.

The first day was uneventful. The crew went through familiar routines of checking the flight system. The payload bay was turned toward the Sun to expose the ultraviolet and x-ray sensors and the plasma monitor to it.

Early on the second day, the white particles that the crew had reported flying by the window during launch assumed menacing proportions. Lousma noticed dark patches on the nose of the ship. Standing close to the window so that he could see directly in front of it, he discovered that a dozen or more of the low-temperature white tiles were missing or had broken away from the heat shield.

While Mission Control was digesting the implications of this report, another arrived from the Kennedy Space Center. Pieces of high-heat black tiles from the heat shield were found around the launch pad and on the beach.

A more detailed inspection of the nose by the crew, and of launch films, found that 25 of the low-temperature white tiles had been shaken off or broken on the nose near the windshield and another 12 tiles had come off the lower section of the body flap at the underside of the fuselage. The 12 were high-temperature black tiles but were in a noncritical area, the thermal protection engineers said. The nose tiles were also noncritical. Heat loads on the nose could be reduced by increasing the angle of attack from 40° to 45° during entry.

Postflight inspection revealed that in addition to 37 tiles lost during ascent, 144 were damaged. These were replaced with denser, stronger tiles, and an additional 852 tiles were replaced or strengthened—a total of 1,033 out of 31,000.

The loss of the nose tiles did not overly concern the crew. In fact, Lousma opened his report with a complaint that he had been awakened at intervals during the sleep period by static in his headset every time *Columbia* passed over the northern latitudes in its orbit.

He had slept in the ejection seat with the headset on while Fullerton stretched out on the deck, with his head braced in one corner and his feet braced in slots of the ejection seat. Fullerton said his problem sleeping in low gravity was what to do with his head. A sleeping bag had been rigged on the mid-deck, but during the early part of the mission both astronauts preferred to sleep on the flight deck.

Lousma wore the headset because mission rules required that one crew member be linked to Houston all the time. The flight deck loudspeaker had failed.

Normally, Lousma should have been left in peace, because Houston would not signal the ship during the sleep period unless Control detected an emergency. But something kept buzzing in Lousma's ear at the northern peak of the orbit.

"The only way I could get rid of it was turn my audio panel off, which I didn't want to do," he said. "But I got a hunch what that is." His hunch was that it was radar tracking the ship as it passed over Iran and within range of the southern Soviet Union.

Neither astronaut was feeling top of the morning. Lousma had suffered nausea and vomiting. The anti—motion sickness prescription of scopolamine and Dexedrine that Fullerton had taken killed his appetite. Still, the crewmen maintained the tradition of sounding good on the air, no matter how they felt.

Lousma observed that from *Columbia*'s 150-mile altitude, it was harder to tell where the ship was at any particular time than it had been on *Skylab*, at 285 miles. "You're so darn much closer to everything," he explained. "I hope everybody on the team is enjoying doing this for real as much as we are."

After the flight, Lousma and Fullerton admitted that day 2 had been their worst day. It felt like the morning after a big night.

During most of the day, *Columbia* flew with its tail toward the Sun so that the payload bay instruments could look away from the sun at deep space. Toward noon, the crew discovered that two of the ship's television cameras were not operating. One was a black-and-white camera at the aft end of the cargo bay. The other was the wrist camera on the manipulator arm.

Both cameras were to be used in the process of maneuvering the arm to grapple the plasma and induced contamination monitors. They were necessary to show the crewman operating the arm from a mid-deck console where to lower the grappling hand, or "end effector."

"There was no picture on the RMS wrist camera when Gordo [Fullerton] tried the end-effector grapple test," Lousma reported to the capsule communicator, Sally Ride. "That's a bad one to lose."

The grappling device consisted of wires that would slip over a knob on the grappling pin of each experiment to be hoisted. With the wrist camera out of order and one of the two aft cameras not working, the crew would have to steer the arm by looking out the rear window. This procedure was complicated by moisture building up on the inside of the after-deck windows.

Control promised to work out a grappling procedure without the wrist camera and also to figure out a way of funneling air across the rear window to keep it moisture free. Testing the arm, the crew reported it worked perfectly. Its stability was not affected when the steering thrusters were fired.

During a pass across the United States, Lousma looked down at California and noticed it was a clear day at Edwards. Any chance it might be dry enough for a landing? Very little chance, Houston advised. Later, over Australia, Fullerton made a video record of tiny white particles streaming from the main engines. These were akin to the famous "space fireflies" first seen by John Glenn 20 years earlier. They were tiny globules of liquid oxygen remaining in the engines.

The wrist camera problem, apparently a short circuit, was not confined to that instrument but also cut out a second camera on the arm's elbow. Both cameras were controlled by the same circuit breaker, and each effort to use the wrist camera caused the circuit breaker to pop, shutting down the elbow camera as well. After discussion with Houston, the crew reset the circuit breaker with the wrist camera off and then turned on the elbow camera. It worked!

Having won this round, Fullerton swung the arm around over the cockpit so that the elbow camera should show the condition of the nose tiles. The extent of the damage was shocking, but it was adjudged not as bad as it looked.

The nose was not subject to entry heating higher than 400° to 600° F. during entry. Even where tiles were broken or missing, the aluminum skin was shielded by a heat-resistant felt padding to which the tiles had been bonded. CC Sally Ride spoke up from Control:

CC: Roger, Gordon. We got a good picture of the missing tiles.
SC: I call your attention to the particles. We're leaving a good trail so we can find the way home.
CC: We were wondering how you navigated.
SC: Anything for show biz, Sally.

During an 11-minute pass over the continental United States, the crew discussed the sleep period noise in Lousma's headset. Control asked if it was coming over the UHF (ultrahigh frequency) channel. That would indicate routine air traffic communication below.

SC: We turned off the air-to-air UHF on the overhead panel after transmitting. Even after I [Lousma] turned it off, I still had some kind of noise coming over the headset when I was at the higher latitudes going over south of the Soviet Union.
CC: Was Gordon bothered by that?
SC: (Fullerton) No. It was very weak because I did not have a headset on. I had the speaker turned full off.

Houston advised that if the noise recurred, it should be recorded for later analysis. The subject was to come up again during the flight.

At suppertime of day 2, the crew inspected the condition of the payload doors after flying all day with the tail toward the Sun. In that attitude, the doors had been in shadow. Had the cold affected them? It had. The crew reported that the aft latches on the port door did not close. Control guessed that the latches had been frozen in the long "cold soak."

SC: Okay, Sally, got you loud and clear. I drove that motor about a minute and 20 seconds or so and I still have no close.
CC: Reopen the starboard door. You may be operating on a single motor.

SC: Gordo is opening the right [starboard] door. Do you want us to open the left door now?

CC: Affirmative. Talk to you in Hawaii in 44 minutes.

Control instructed the crew to roll the ship to top Sun attitude so that the Sun would shine directly on the payload doors as soon as the ship entered the daylight portion of the orbit. The crew changed the ship's Sun attitude as recommended, and *Columbia* entered sunrise as it passed within communication range of the Seychelles tracking station in the Indian Ocean. In a few minutes, the frozen latch mechanism thawed and the latches worked properly. Something new had been learned.

As the crew prepared for sleep, conversation with Mission Control turned to the mysterious night noises.

SC: (Lousma) I noticed last night that beside a very few weak UHF transmissions that I wasn't able to get rid of the noise by turning off the UHF. And it sounded like when somebody's tracking you with radar and it seemed like it got into us without going through UHF.

CC: We understand that, Jack, and we're trying to record it. As far as the tile goes, we think there is no concern because of the tiles that are missing. We've made an assessment using previous flight data and we think that the maximum structural temperatures aren't high enough to compromise the strength and integrity of the orbiter structure.

SC: Okay. If you find out otherwise, I don't think we want to know.

Through the tracking station at Santiago, Chile, Mission Control finally received a report on the payload doors.

SC: Hello, there, through Santiago. Gordo's got some good news on the door business. It worked all right.

CC: Super. We're all very glad to hear that down here.

SC: You're not any gladder than we are, brother. We started the door closed . . . and we got the latches all back together in 29 seconds.

Astronaut David Griggs, the capsule communicator on the late shift at Houston, advised the crew to "sleep in tomorrow." Private medical conferences between the crew and flight surgeons had dealt with gastrointestinal distress and lack of a good night's sleep. Lousma particularly seemed overly tired.

SC: (Lousma) Okay. Well, if we get to feeling better, we'll be ready to charge. Actually, we weren't overpressed today, I don't think. We had all the food we wanted. I don't know—it's pretty hard to sleep up here.

CC: The plan is—you just give us a call when you're ready. Fuel cell purge needs to be done tonight.

SC: I guess the thing most bothersome for me last night was that I was sleeping with an earphone on and every . . . there would be some spurious noise when we were at our high latitudes over Iran and around in China and in there. If there was a way to avoid it, that would be good. It wasn't any UHF. Somehow there's something sneaking in there that's making it into my headset. [It] had the same character as when you're driving along in an airplane and somebody paints you with radar.

The Electron Gun

The third day opened with a cheerful voice from *Columbia*:

"Good morning, Dave," said Lousma. "Top of the morning to you."

CC: We're listening.

SC: Okay. We slept a little better last night and we've been up and around a little bit. I presume we ought to get in nose-to-Sun attitude pretty soon.

It was 8:20 A.M. eastern standard time. Mission Control played a recording of the Marine Corps hymn.

SC: (Lousma) I'm standing up, by the way.

CC: Which way is up, Jack?

Another malfunction had appeared. The toilet was stopped up. Space toilets are a new invention, and this one had its share of complexity. A motor-driven device called a "slinger" separated feces from urine for treat-, ment and storage. The motor had popped a circuit breaker and stopped. Something seemed to be impeding the slinger.

Control advised the crew to reset the commode circuit breaker. The crew did so.

SC: It worked a while and then stopped. What makes it pop is the slinger is loaded up and is not turning real fast and pretty soon it just pops the breaker.

After a brief analysis of the problem, Control concluded that the slinger was caught on a waste bag. It would continue to operate at low speed by dragging the bag around with it. But at high speed, it would jam and pop the breaker. Control's remedy: run it at slow speed.

During the day, the crew turned on the Vehicle Charging and Potential Experiment. It monitored the orbiter's electrical characteristics and telemetered the data through the ground stations to Houston. The electron beam gun was turned on. It emitted a pulsed beam of 1,000 electron volts at a current of 100 microamperes.

In theory, the beam would be guided by the Earth's magnetic field and would spiral around lines of magnetic force as it left the cargo bay. It would emit a faint light as it passed through the electrified plasma.

Fullerton reported that a filament of light from the beam could be seen in a televised image. It was a straight streak angling out at 45° from the bay. But the streak was faint, probably because the camera lens recorded it through a fogged rear window.

Another problem appeared, noted this time at Mission Control. Its telemetry data showed a 20-pound drop in the pressure of a nitrogen tank that could not be accounted for by use. The storage tank was one of two that provided nitrogen for the cabin atmosphere.

On board, the crew had not noticed any change in the cabin air. "Okay," remarked Lousma. "We haven't been breathing hard."

One of the tanks was leaking. It was not alarming, but it was nagging. Eventually the loss of nitrogen was traced to tank number 2, and it was shut off.

In the matter of the fogged rear window, Control had worked out a fix: Funnel air flow on the window. Use a sheet of plastic to make the funnel. Where could plastic be found aboard *Columbia?* Control suggested that the back of the World Atlas was plastic. Rip it off. It would do. And it did.

The crew turned on the electrophoresis experiment, which ran itself once started. Blood cells were migrating from serum into glass tubes under the gravity-free attraction of an electric field. Gravity interfered with this method of separation on the ground.

The plant growth experiment was going well. The question it was designed to answer: Would lignin grow normally in low gravity? Lignin is the stiffening agent in plant stems. It holds them up against gravity. Fullerton reported that the seedlings were growing.

CC: Fine. Keep the fruit fly away from the mung beans.
SC: I haven't seen him. I think he finally found a place, because he was getting pretty good at flying around here.

Lousma cited another contrast between observations from the low altitude of *Columbia* and from the higher one of *Skylab.*

"You can really see a lot of detail from this altitude compared to *Skylab,*" he said. "In *Skylab,* you could only see freeways if there was some contrast. Here you can see minor roads even, and you can see small settlements and when you get a building big enough where you can see—you can see it with the naked eye. But it's a lot harder to tell where you are, though, because you're so much closer."

CC: You should be passing the East Coast right now.
SC: Yes, we just passed the coast, which is totally cloudy.
CC: Jack, I've got our latest thinking on the WCS [waste control system] if you want to listen.
SC: Yes. I'll probably get more and more interested in that as time goes on.
CC: There might be a bag stuck down in the bottom which is putting a drag on the motor. The system should work in the emesis [slow] mode.
SC: Okay. Thank you.
CC: Jack, one more thing. If the slinger does stop completely, we'd like to know about it. We have a couple of more tricks up our sleeve that we could try.

SC: That's kind of a bad choice of words because if you want to try 'em it means I'll probably be up to my sleeves.

CC: Glad to hear you're feeling better, Jack.

As *Columbia* once more crossed the California coast, the crew turned on the television camera to display the scene below.

SC: We went over the Golden Gate a couple of minutes ago. Looking at Lake Tahoe now. You're seeing the mountains through Nevada with snow on most of them. You should be getting the scene from Delta camera. . . ."

CC: We can see both wings in the picture. That's a pretty machine you're flying.

SC: It is that. It really flies smooth. Now we're coming over midwestern farmland. Looks like a patchwork quilt. Farmers have all their fields in different states of cultivation or growth and some are dark green and some others are very light brown and lots in between. It really is America the beautiful.

CC: Roger, Jack.

SC: We can even see some airplane contrails down there. Looks like they're heading for Chicago. Looks like they've got a lot more weather over the eastern part of the U.S. today.

Heading down the Atlantic Missile Range, Lousma put the ship into the barbecue roll and prepared to direct an inside *Columbia* television show, starring himself and the bees and the moths. The show began on the next pass over the United States.

"Hello, there, space fans," he said. "Here we are in the good ship *Columbia* speeding over the United States."

The camera showed a cluster of bees huddled sorrowfully against the glass of the cage, and next to them some moths, flitting about. The hitchhiking fruit fly had taken cover.

As the crew heated up some supper, the capsule communicator, David Griggs, called up a congratulatory message:

"At 2 days, 6 hours, 20 minutes mission elapsed time, you became the holders of the Shuttle record of 2 days, 6 hours, and 21 minutes."

On day 4, the crew had regained the easy confidence and good spirits of veterans. Fullerton went to the mid-deck RMS console and exercised the arm. Carefully, he maneuvered it so that the end effector would grapple the knob of the grappling pin on the plasma diagnostics package (PDP). Suddenly, he latched onto it and lifted it clear of the pallet.

Lousma proudly displayed the scene via the television camera as *Columbia* came within range of MILA, the Kennedy Space Center tracking station.

SC: Okay. Gordo's picked up the PDP. You can see it there on the arm. It's sort of resting there above the payload bay.

The arm appeared over the open payload bay, swinging a cylinder. Fullerton lifted it high over the bay and tried to swing it through the beam from the electron gun.

He extended the arm far forward over the nose of the ship. The elbow camera showed the broken and missing tiles in detail. Damage spread across the entire nose of the ship. It had certainly taken a beating at launch, probably from ice falling off the external tank. It looked as though it had been pummeled by a shower of meteorites.

The plasma diagnostic package, containing particle and radiation sensors, is shown extended over the side of *Columbia*'s cargo bay by the remote manipulator arm. Fullerton took the picture through the ceiling window.   (NASA)

Throughout the mission, Lousma remained fascinated by the view below. During day 4, as the ship approached the night side, he waxed lyrical about it: "We're moving into the darkness, and the edge, the rim of the Earth is very poorly defined. It goes gradually from very dark blue and fades into black, but there is no real definition; it's kind of like you're going out into nowhere and it's one of the most interesting contrasts I've seen. However, we can see the Sun shining on the . . . buildup of clouds on the other side, kind of reddish in color, but we are finding it's not very well defined; it is a very poor gradation between what you see on the Earth and then the blackness of space; it's kind of, I guess, a lonely feeling or a feeling of not knowing exactly where you're going because you can't define the edge of the Earth or see any stars yet."

"Sounds like a beautiful sight, Jack." It was Sally Ride speaking. A year later, she would see it, too, on STS-7 (June 18–24, 1983) aboard *Columbia*'s twin ship, *Challenger*.

At suppertime, the crew re-berthed the plasma monitor, lowered the arm into its cradle, and shut off the electron beam. During the day, experiment data, voice, and TV had been transmitted in the high-powered, high-rate channel of the ship-to-ground, phase modulated S-band radio, the primary communication system. To conserve power, the crew switched the ship-to-ground, or downlink, transmission to the low-powered, low-rate channel. At this point, three of the four downlink channels of the primary system failed.

The system operates through two onboard transponders. Only one works at a time. The other is a backup. Each has two downlink channels, one high, one low. When the crew discovered that the low-powered channel on the transponder it was using was out, it switched to transponder 2, only to find that both its high- and low-powered channels were dead.

The only downlink channel still working in the primary system was the one the crew had been using all day.

Mission Control was not only puzzled by this unexpected breakdown, it was flabbergasted. Three-quarters of the phase modulated S-band transmission had mys-

teriously died. It had never happened before. Curiously enough, the uplink channels in the system remained unaffected. The trouble seemed to be localized in the ship's two transponders. Technicians at Houston suspected that something had happened to a logic circuit.

In addition to the phase modulated (PM) channels, the S-band system has frequency modulated (FM) transmission. However, the FM signals are detoured through onboard recorders, and voice communication by FM would have to be played back by recorders unless they were rewired. The rewiring could be done by the crew with instructions from the ground, and Control considered that option.

After hours of analysis, Mission Control decided to leave well enough alone as long as the high-powered PM downlink channel was functioning. There was fear that tinkering with the system might kill that channel.

Reporters spent hours in briefings and conferences with officials trying to find out whether the failure of three of the four main downlink channels would require the directorate to terminate the flight. No, replied the directorate. There was plenty of FM and UHF backup.

Did anyone care to speculate on what had happened to cause the worst communications failure in the history of U.S. manned spaceflight? No one did.

Had the electron gun beam "zapped" the S-band transponders? Not a chance, insisted the communications people at Houston. There was no way that the electrical properties of the transponders could have been affected by the stream of electrons emitted by the gun, according to the deputy director of flight operations at Houston, Eugene F. Kranz.

Yet the failure has not been accounted for publicly at this writing, and the equipment was replaced by the contractor, TRW, Inc., for *Columbia*'s fourth and final test flight.*

Day 5 of the mission found the crew cheerful and not much concerned about the communications problem. It appeared to some of us who have observed manned spaceflights since they started that an extraordinary number of failures had occurred on STS-3: In addition to

---

*NASA said in response to my query that the replacement parts were relays.

losing three of four S-band main downlink channels, the ship had lost at least 37 tiles from the heat shield; one of two nitrogen storage tanks was leaking; one of three auxiliary power units had overheated during ascent, and no one could be sure if the problem would recur on descent; two television cameras had failed; and the toilet was out of order. None of these malfunctions was regarded as serious enough to curtail the flight. Consequently, Houston remained calm and the crew cheerful. Once more, the arm picked up the plasma package.

SC: Got an interesting picture looking out the overhead windows. We're upside down, trailing over lots of clouds. Black sky is in the background and the PDP is leading the whole show. We've got the arm way up over the cockpit with the PDP hanging up over the nose.

Lousma observed that it was a cloudy Friday morning (March 26) over the west coast.

CC: Looks like you guys will probably fly for a while longer, huh?
SC: (Lousma) That's right. Got any weather forecast for White Sands for Monday?
CC: We believe it's going to be good for Monday, Jack.
SC: I see, Steve, that we're flying directly over the Marine Corps Air Station at Cherry Point, North Carolina . . . the home of a number of famous Marine aviators.
CC: Roger.
SC: (Fullerton) Yeah, both of them live there.
SC: (Lousma) I won't dignify that with a reply.

The crewmen carried small bar magnets in their pockets. When they let the magnets float about in the cabin, the bars immediately lined up along the magnetic field of the Earth, like a compass needle. The bar magnets showed how the long axis of the ship was aligned with the direction of the field. Any change in the ship's attitude would be promptly shown by a change in the position of the magnets.

On the 68th orbit, capsule communicator Brewster Shaw called up with news from home.

CC: We ran into Gratia [Mrs. Lousma] and Marie [Mrs. Fullerton] in Building 30 here and after they heard you were feeling better they were all smiles and really enthusiastic about the great job you are doing.
SC: Well, great. Glad to hear that.
SC: I'll say that. It's lonesome up here but we'll be back this Monday and tell them all about it.
SC: Well, I'll have to be home Monday because I'm running out of clean underwear.

Fullerton asked Brewster to convey wishes for good luck to young Andy Fullerton, age 6, for Little League tryouts Saturday.

Control then passed up word that attempts to troubleshoot the S-band transponder failure had not produced any results. Meanwhile, at the insistence of news correspondents, flight director Neil Hutchinson attempted to clarify the state of the crew's health at the change of shift briefing. Like a suspect being interrogated by a grand jury, he admitted, yes, Lousma had vomited once. He said that Dr. Sam Pool, one of the flight surgeons, had prescribed an antacid for Fullerton, who had complained of stomach gas. The physician had diagnosed too much fluid intake and not enough food during the early days of the flight as the cause of Fullerton's problem. For example, one morning Fullerton had eaten some granola with blueberries and had drunk grapefruit juice, for a total calorie intake of 400. That was not enough, Dr. Pool said.

"We do not think they've been eating, and they say they haven't been eating their entire meals," he explained.

He attributed lack of appetite to the anti–motion sickness medication. But its effects were wearing off, Dr. Pool added, and on the fifth flight day, the crew would feel more chipper. He was right.

At the end of day 5, Control advised the crew that day 6 would be theirs to do as they liked. The experiments were working and the arm was performing beautifully. The only part of the flight plan to be omitted was lifting the Induced Environmental Contamination Monitor out of the pallet. The exercise was cancelled because of the difficulty of grappling the experiment without the wrist camera. It was not in as available a position on the pallet as the plasma monitor had been.

During day 5, *Columbia* exceeded the combined flight

times of the first two missions—108 hours—at 11:35 P.M. Friday, March 26.

Down Home

The sixth day began at 5 A.M. EST. It was to be a day of getting the ship ready for a Monday entry.

CC: Nothing for you this morning except some sounds from your families if you're ready to listen.
SC: Ready.

Up came the recorded voice of Timothy Lousma, 18. "All right, you guys! Wake up! Five more minutes. Better get up!"

From Mary Lousma, 13: "How'd you sleep last night, Dad? I sleep with my eyes closed." (A family joke.)

A word from 18-month-old Joseph, coached by mother:

Gratia Lousma: Can Joseph say "Daddy"?
Joseph: Daddy.
Gratia: What does a cow say?
Joseph: "Moo."
Gratia: Good morning, honey. I'm sure all that you've just heard brings you back to Earth again. We don't want you to get too comfortable up there and decide to stay!

And then the Fullertons.

Andy, 6: "Good morning, Dad. Don't forget to look out and see Texas. I'll be waving at you."
Molly Marie Fullerton, 8: "Good morning. Time to get up and I hope to see you at the landing. Love you and miss you."
Marie: "Morning, Gordo. This is Marie. Mom. We love you. Have a good day, Gordon and Jack."
SC: Well, it sounds just like home. It was beautiful, sunny as the alarm went off, and we pulled up the window shade and looked out there. And about halfway through the pass the sun went down and now it's black as night.
CC: Roger.
SC: I guess we'll just roll over and go back to sleep.

Instead, after breakfast over the Mediterranean Sea, they worked out on a treadmill designed for keeping muscles in shape by Dr. Jack Thornton, a physician-astronaut.

Landing was in the offing now. Check tire pressure, Control said.

Wave-Off

Another day, another malfunction. On day 7, the cathode-ray tube display in front of Lousma's seat failed. Sally Ride advised that the problem might be in a key that could be changed. A key could be taken out of a display unit downstairs and substituted for the defective one. The fix worked.

Two major events came to light. The solar x-ray monitor had detected powerful signals from a new solar flare. APU 3—the one that had overheated during ascent—was started up briefly. The cooling system seemed to be working normally.

During the evening, the crew put the ship into the barbecue roll and prepared for Monday morning entry. Landing was scheduled at White Sands at 2:27 P.M. EST Monday, March 29, or mission elapsed time of 7 days, 3 hours, 27 minutes.

Control awakened the crew at 5 A.M. By 7:45 A.M., the

C. Gordon Fullerton, STS-3 pilot, struggles into a modified, high-altitude Air Force pressure suit prior to landing. (NASA)

payload bay door radiators had been retracted and the doors had been closed. Messages of appreciation were radioed to the crew from controllers, experimenters, and management officials. While Fullerton struggled into his flight suit, Lousma called for a weather report and a recommendation for runway selection at White Sands.

CC: We're anticipating a right turn into 17, but the winds are pretty gusty out of the south and we're checking it for you.

Runway 17 was the one that had been equipped with the microwave beacon that would guide the autopilot to landing. The flight plan called for the crew to turn control of the ship over to the autoland (automated landing) system at 12,000 feet on the final approach. At 200 feet, the crew would resume manual control and drop the landing gear. This procedure would partially test the autoland system, which NASA hoped to use on all operational flights.

Two hours and 56 minutes before the reentry maneuver, John Young took to the air at White Sands in NASA 946, a Grumman jet trainer, to have a look at the weather. He reported severe turbulence aloft, winds blowing at 40 to 56 knots at 2,000 feet altitude. Houston was dismayed.

Columbia passed over the United States, and Fullerton reported that he was all suited up, while Lousma was getting his harness on. Shuttle crews don a stretch undergarment for landing to prevent blood from pooling in the lower abdomen and legs under the influence of full gravity. Fullerton added cheerfully that he and Lousma were having a snack.

CC: The weather is clear and good visibility at Northrup. Forecast winds on the surface are to pick up, so we're going to have to watch those all the way to deorbit burn. And there's some probability of a wave-off if the winds go out of limit. But we don't anticipate that at this time.

There had never been a wave-off in manned spaceflight. It means unsafe conditions for landing, and it instructs the pilot to go around again or remain aloft until conditions are safe. In the case of an aircraft, a wave-off usually means circling the field. In the case of *Columbia*, it meant going around the Earth.

Columbia was expected to land on orbit 116. The ground track passed through the Los Angeles metropolitan area, which *Columbia* would overfly at 100,000 feet as it descended toward Arizona and New Mexico.

On orbit 114, Control advised the crew, "You have a 'Go' for Ops 3 transition." Ops 3 was the descent and landing program for the flight computer.

CC: *Columbia,* Houston. One note of interest. The burn TIG [time of ignition] will occur on the next time around about 18 seconds after AOS [acquisition of signal] at Yarragadee.

SC: Okay, thank you. How're the winds and so forth holding at Northrup?

CC: Well . . . so far, so good. I hope to have more words for you at Hawaii.

Meantime, John Young, scouting the weather at White Sands, called Houston: "Winds aloft—pretty brisk."

CC: Joe has a few words to pass along about final approach and landing.

Control turned the mike over to Joe Engle. He recommended a procedure for turning into runway 17. But worsening weather was worrying everyone.

John Young reported that a virtual dust storm had started blowing over Northrup Strip. Winds were gusting up to 36 knots and visibility was deteriorating.

"I think you ought to knock it off," he recommended.

"Concur," said Mission Control, Houston.

That exchange made manned spaceflight history. For the first time, a spacecraft was waved off a landing. Word of the wave-off was passed to the crew 30 minutes before deorbit time.

Columbia's flight was extended one day—to eight days. Deputy Flight Director Kranz assured the news media correspondents that there were plenty of consumables—fuel, air, and water—on board to extend the flight two days longer than that.

However, the flight directorate considered it prudent

to bring the ship down the next day, Tuesday, March 30.* If the weather did not clear up at White Sands, the crew would be authorized to land at the Kennedy Space Center.

"We will have targeting available that could allow us to deorbit to Kennedy," Kranz said.

The decision sent a shock wave of surprise through all the NASA centers, and Kennedy instantly mobilized its troops to receive *Columbia* if, indeed, it had to land there six months ahead of schedule. As word spread through Brevard County, Florida, thousands of motorists began parking along the causeways to wait for the possible arrival of *Columbia* from space.

Whatever the weather was at White Sands, it had been dreadful at Kennedy for two days. Wind-driven rain had been beating down on the Kennedy Space Center for hours at a stretch and there were fears of tornadoes. The weather was so bad Monday afternoon, March 29, that the prospect of a Tuesday morning landing anywhere in central Florida seemed highly imaginative.

But weather forecasters insisted that the winds would die away, the rains would stop, and the sun would shine in the magical way bad weather suddenly yields to fair winds and sunshine in Florida. This would come to pass by noon Tuesday.

As for White Sands, there was a possibility of clearing there, too, if a high pressure system in the Gulf of California moved eastward into New Mexico.

Houston made it firm that White Sands was first target of landing opportunity and Kennedy second.

CABU to Mars

Lousma and Fullerton repeated Sundays closeout preparations on Monday and then retired at 7 P.M. Houston woke them at 3 A.M. Tuesday. White Sands was clearing.

CC: You might be thinking about runway 17 options.
SC: Then we might also be thinking about right turns.

*Presumably in view of the loss of three S-band downlinks.

As *Columbia* approached White Sands, it would turn in a wide loop, the heading alignment circle (HAC) to align itself with the runway. The turn would be made to the right.

Control passed up deorbit burn instructions. At White Sands, the magic was working. Skies were clearing, winds dying. Again flying weather reconnaissance, John Young reported:

"Right now it looks like a right-hand turn into 17. The visibility is CABU [clear and beautiful] to Mars. You can see all the way to the San Francisco Peaks, and it's clear as far west as we could see."

CC: *Columbia*, Houston. You are Go for Ops 3 transition.

Once again, the crew set the flight computers for the entry and landing program. Crossing the equator as it flew northeastward across the Pacific, *Columbia* began orbit 128. The crew tested all three APUs and verified that they were working. Power steering would be turned on next orbit for descent and landing.

On the last MILA pass, Houston called: "Okay, Jack, you are Go for deorbit burn."

SC: We're getting suited up here and getting our hats on.
CC: Have a good burn.

When the ship passed within range of Yarragadee, Australia, the crew called:

"Ignition, Houston! We got a good burn going."

The twin 6,000-pound-thrust maneuvering engines were fired in the direction of flight for 2 minutes, 29 seconds. They cut velocity by 286.2 feet per second and altered the shape of the orbit so that at perigee it intersected the Earth. In fact, its low point was 19 nautical miles underground!

CC: *Columbia*, weather at Northrup is excellent. Surface winds are nearly calm at this point. Next, Orroral in two minutes,
SC: We got you. All clear.

Columbia was descending toward the atmosphere, where entry would begin at 400,000 feet. The crew had turned the ship around so that it descended nose up toward its encounter with the shell of gases that enfolds the Earth. It was now a glider.

CC: (through Orroral Valley station, Australia) Roger, *Columbia*. Houston. You are Go for maneuvers [to prepare for entry]. A reminder to close the vent doors and remain configured to AOS through entry. We're 30 seconds to LOS [loss of signal]. We may pick you up a bit in Hawaii. If not, we'll see you in 23 minutes over the States.

Columbia's ground track crossed the coast of Baja California, Mexico, about 50 miles south of Ensenada. It went northeast across Mexico and entered Arizona between Yuma and Lakeville. It crossed the state line into New Mexico north of Silver City and passed near Truth or Consequences just before descending into White Sands.

CC: *Columbia*, Houston. Through Hawaii.

No answer.

CC: *Columbia*, Houston. Through Hawaii. Over.
SC: Okay. We're hearing you through Hawaii. All is well.

C-band radar in California picked up *Columbia* 847 miles out. The ship was descending rapidly.

CC: (through the Buckhorn station, California) Energy and ground track good and NAV [navigation] is great, Jack.
SC: That's good news. Thanks, Steve [Nagel]. This is really a beautiful flying machine.
CC: That's great to hear, Jack. We show you passing Mach 14 at 179,000 feet.
SC: That's affirmative. We got the coast of California in sight and we're about to go over L.A. in about Mach 16, correction, Mach 13. I think we're booming right over the Commander-in-Chief's (President Reagan's) ranch right now, Steve.

This may have been a slight exaggeration. The ground track on this orbit was considerably south of the Los Angeles area, but Lousma could have had it well in view from his left-hand seat.

The speed brakes on the rudder opened at 167,000 feet altitude and 440 miles from Northrup.

PAO: Three hundred and fifty miles to White Sands; 150,000 feet.
SC: Okay. Looks like we're going right over Phoenix at Mach 9. We're passing Davis Mountain.
PAO: 124,000 feet. Mach 6.2, range 186 miles. Approaching state line between Arizona and New Mexico.
CC: Passing 90,000 [feet]. Positive seats. [This meant the crew could safely eject if necessary.]
SC: A little bumpy around Mach 2.

Columbia appeared on the television monitor, flying like a sparrow, a seagull, a hawk, an eagle. Rapidly it grew into a big flying machine.

CC: Columbia intersects the HAC now, passing 26,000 feet.

Columbia lands on the Northrup Strip of the White Sands, New Mexico, missile range at the end of its eight-day mission.

(NASA)

Lousma made the much-discussed right-hand turn into runway 17.

CC: *Columbia*, Houston. Go for auto [autoland] to enter glide.

Lousma relinquished control of the ship to the autoland system as the ship descended through 12,000 feet, 2½ minutes from touchdown.

PAO: Five thousand feet. Airspeed 280. Range, three miles.
SC: Preflare now.

Columbia levelled off, its speed dropping.

PAO: One thousand. Still in auto.
CC: Fifty feet. Keep coming.

PAO: Gear down. Twenty feet, ten, five, four, touchdown.

Columbia rolled majestically down the runway. Fearing that the nose wheel would come down too hard and too fast, Lousma pulled the nose up slightly to retard the letdown. The action caused the nose to lift momentarily as though hit by a gust of wind. Then it dropped and the nose wheel settled to the ground.

PAO: Mission elapsed time is 8 days, 4 minutes, 49 seconds.

It was 11:05 A.M. March 30, 1982.

SC: Wheels are stopped, Houston.
CC: Okay, *Columbia*. Welcome home. That was a beautiful job.

The Fourth Test

Eighty-nine days after its third landing, *Columbia* was launched due east on June 27, 1982, on its fourth and final orbital test. The mission was scheduled for seven days. The orbiter would land at the Dryden Flight Research Center, Edwards Air Force Base, in the desert July 4.

Liftoff from Kennedy Space Center was on time—for the first time—at 11 A.M. eastern daylight time (actually a few milliseconds before the hour). The launch seemed to go smoothly enough until reports filtered into KSC and the Johnson Center that after being cast off, the solid rocket boosters had been lost in the Atlantic Ocean 160 miles off the Florida coast. Instead of floating as before, they sank before recovery ships could reach them.

Moreover, the performance of the boosters was lower than expected, causing the ascent trajectory to be lower and slower than planned.

Mission Control called the crew's attention to the depressed trajectory at 2 minutes, 35 seconds into the flight.

It altered the timing of abort procedures in the event of main engine failure. However, the crew had already noticed the booster performance discrepancy and was aware that the main engines and the orbital maneuvering system could easily pick up the slack.

About 18 minutes after liftoff, a leak was found in the forward manifold of the reaction control system. It was not serious enough to have an effect on the system's overall functioning, however, and consequently was not a constraint on the flight.

Despite these and other minor malfunctions, *Columbia* reached its cruising altitude of 160 nautical or 184 statute miles on time with four firings of the orbital maneuvering engines. Mission Control sent a "Go" for on-orbit operations, and the flight commander, Thomas K. (Ken) Mattingly II, 46, a Navy captain, responded, "All right, sir. In that case, we'll get out of our formal go-for-orbit clothes and get into something more comfortable." He and the pilot, Henry W. Hartsfield, 48, who retired

Thomas K. (Ken) Mattingly II, commander (right), and Henry W. Hartsfield, Jr., pilot, of the fourth voyage of *Columbia*, STS-4, arrive at Launch Pad 39A for the preflight countdown demonstration test. In this test, crew procedures during the countdown are rehearsed. The crewmen are accompanied by Jean String, a KSC security specialist.   (NASA)

from the Air Force and joined NASA when the Manned Orbital Laboratory program was cancelled, then struggled out of their pressure suits and put on blue NASA coveralls.

Mattingly had flown to the Moon in 1972 on *Apollo 16* with John Young and Charles M. Duke, Jr. While they were exploring the Descartes highlands, Mattingly had remained aboard the command module in lunar orbit making radiometric surveys of the surface. These turned out to be immensely valuable in revealing the mineral wealth of the Moon. Hartsfield was in orbit for the first time.

Before the first orbit was completed, it was evident that *Columbia* had reached a level of maturity that made its space operations appear routine, despite minor failures. The turnaround time from STS-3 had been reduced by 34 days, that is, from 114 days to prepare the third flight for launch to 80 days to get the fourth flight ready at KSC.* The loss of the boosters in the Atlantic was costly. The cause was diagnosed as a switch malfunction that

---

*Turnaround time does not include processing time at Edwards AFB and transit time to Kennedy Space Center.

released the three big parachutes on each booster before it reached the water, and not a design flaw.

In all major respects, the Shuttle was working as it was designed to do, and the great unknowns—the performance of a winged spacecraft during entry into the atmosphere, the validity of the ceramic heat shield, and the aerospace craft's handling in the atmosphere as an unpowered glider—had been resolved. Unless some unexpected failure or crippling effect occurred on the fourth mission, NASA's Space Shuttle Transportation System would be certified by the agency as operational.

Columbia had already demonstrated its utility as a platform for scientific observations on the second and third missions. In revealing the existence of prehistoric river systems below the bone-dry sands of the Egyptian Sahara with an experimental radar imaging system, the second mission had scored a genuine breakthrough in the technology of remote sensing from space.

STS-4 carried, in addition to manufacturing and scientific experiments, a secret military payload ensconced in the cargo bay. The Department of Defense cargo was known officially only as DOD 82-1. It was a forerunner of the Defense Department's increasing dependence on the Shuttle in future years.

An intense aura of secrecy surrounded DOD 82-1. The Air Force was adamant in refusing to discuss it, and no one in NASA was allowed to do so. However, if DOD 82-1 was designed as a test of keeping the nature of a payload secret, it failed on that score, for the press was able to publish a detailed description of it.

It appeared that the secrecy had been breached by the Department of Defense itself during NASA authorization hearings before the Subcommittee on Space Science and Applications of the House of Representatives in 1981. An assistant Secretary of the Air Force, Robert J. Hermann, told the subcommittee on March 19 that "we are planning to place a critical space test program experiment called CIRRIS (Cryogenic Infrared Radiation Instrumentation for Shuttle) on the fourth OFT [orbital flight test] mission."[1]

The system would provide critical information applicable to future defense missions, Hermann said. It would give both the Air Force and NASA an early opportunity to evaluate "procedures and interfaces for operation with the Shuttle." He explained that "we have committed all of our operational space programs to an orderly transition to the Space Shuttle."

A key instrument in the CIRRIS package, an infrared telescope, failed to operate because the telescope cover would not open on command from the crew. Mattingly and Hartsfield tried to dislodge the cover by nudging it with the end effector, or hand, of the remote manipulator system arm, but were unable to do so. They were heard to report "no joy" (no luck) in conversations with the Air Force Data Processing Center at Sunnyvale, California.

Inasmuch as a model space suit had been taken on the flight so that Mattingly could try it on for size and fit, someone suggested that the flight plan be changed to include a space walk. Mattingly would don the suit not merely to model it but "for real," slip through the airlock into the open cargo bay, and manually remove the telescope cover.

Without disclosing the source of the idea, Mission Control rejected it. Officially, it was explained that the exploit would throw the tightly controlled mission time line off schedule. Unofficially, headquarters took the position that the fix was hardly worth the risk. A space walk was not planned until STS-5, the next flight, and then it was assigned to two mission specialists, not to the pilots.

Two other instruments in DOD 82-1, said to be an ultraviolet scanner and a sextant, reportedly functioned properly.

STS-4 was distinguished also by carrying the first commercial space manufacturing experiment and the first operational Getaway Special. The "Special" is a program for flying small, individual, self-contained, and self-operating experiments aboard the orbiter at nominal cost for small businesses, laboratories, universities, and private individuals.

The commercial experiment consisted of an engineering model of the continuous-flow electrophoresis system. The hardware, as mentioned previously, was provided by McDonnell-Douglas Astronautics Co. in cooperation with the Ortho-Pharmaceutical Division of Johnson & Johnson. The test had been arranged under a joint endeavor agreement between NASA and the companies providing for six free flight tests of the equipment. If it proved commercially useful, future flights would be charged a scheduled transportation fee.

In the joint endeavor program, NASA subsidized equipment development by providing a number of free rides. After that, users would pay.

As I mentioned earlier, electrophoresis is a method of separating biological materials by their suface electrical charge as they move through an electrical field. On the ground, the process is impeded by convection arising from electrical process heat. In orbital free fall, convection, which is produced by gravitation, is absent. The microgravity environment of orbit also makes it possible to process larger samples to high purity than presently seems possible on the ground.

A number of scientists are not sold on electrophoresis in space as a manufacturing option because they think the interference of 1-G can be nullified in some way other than by costly spaceflight. The folks at Johnson & Johnson and McDonnell-Douglas Astronautics obviously did not share this view. They invested in a Continuous Flow Electrophoresis System (CFES), which NASA agreed to test on STS-4. Earlier electrophoresis testing had been conducting in 1975 on Apollo during the Apollo-Soyuz Test Project, and the electrophoresis experiment on STS-3 used that equipment. The new machinery for STS-4 was expensive looking.

The STS-4 testing was designed to separate enzymes and hormones from cultures of human cells obtained from cadavers. It was hoped by the experimenters that the output of CFES would be 300 to 400 times that of a comparable process on the ground.

As did STS-3, the fourth mission carried another processing experiment, the Monodisperse Latex Reactor. It consisted of four reactors that automatically make tiny latex spheres of the same size (monodisperse), several microns in diameter. The beads are used in medicine and also in industrial research. In medicine, as mentioned earlier, they are deployed through the bloodstream to carry medication or radioactive isotopes to tumor sites. They are also used to determine the size of lesions in the intestine or in the eye. Industrially, they are used in calibrating tolerances.

On the ground, beads were limited in size to 3 microns, according to experimenters. The experimental system was designed to determine if they could be made larger, up to 20 microns, and still be all the same size. Results on STS-3 indicated that larger sizes can be made in orbit. The fourth flight was expected to confirm that finding.

The first operational Getaway Special consisted of nine scientific experiments housed in a cannister about the size of a 30-gallon garbage can. The space on board was purchased for $10,000 by R. Gilbert Moore, manager of the Thiokol Corp., the manufacturer of the Shuttle's solid rocket booster motors. The space was then donated to students at Utah State University.

The experiments consisted of low-gravity tests of fruit fly and shrimp growth; the effects of microgravity on the growth of duckweed roots and one-celled algae (chlorella); surface tension of melted solder under low gravity; the efficiency of low-gravity soldering; alloy formation of tin and bismuth under low gravity; thermal conductivity of an oil-water mixture in orbit; and low-gravity effects on the curing of a composite of epoxy resin and graphite.

During the first part of the mission, the crew was unable to turn on electrical power for the experiments that required it. A search disclosed a defective switch. The crew hot-wired around it and the full complement of experiments became operative.

So delighted were the students at this fix that they sent a message via Mission Control to the crew, paraphrasing Neil Armstrong's "one small step" comment when he stepped on the surface of the moon. "One small switch for NASA," the students said, "a giant turn-on for us."

Hartsfield deployed the 50-foot-long remote manipulator arm to pick up the Induced Environmental Contamination Monitor. Its array of instruments, including a mass spectrometer and photometer scanned the cargo bay for waste gases and particles from the space engines, which might interfere with the scientific instrument data. With a mass of 800 pounds at launch, the monitor was practically weightless in orbit, but its mass provided a test of the arm's maneuverability. Hartsfield swung the instrument array around the 60-foot cargo bay with ease and precision.

Curiously, this exercise ended for a time the use of this

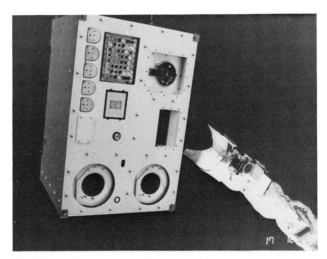

This box, called the Induced Environment Contamination Monitor, contained a mass spectrometer and other instruments that identified gases and particles released by *Columbia*'s space engines and emissions from other sources during flight. Here the box is shown deployed by the remote manipulator arm to monitor these contaminants. In theory, they could affect the performance of other scientific instruments. (NASA)

remarkable Canadian-built feature of the orbiter. Although initially planned as a device to deploy satellites and retrieve them, the arm would not be used to deploy a pair of communications satellites on the first operational mission. Instead, they would be spun out of the cargo bay one at a time by springs from revolving turntables and then boosted by attached rockets to geostationary orbit.

Early in the mission *Columbia* exhibited a tendency to pitch and roll, causing its attitude control jets to fire to stabilize it. These motions were puzzling until it was recalled that the Shuttle had been drenched by a downpour of rain and hail the day before the launch. It was probable that moisture had been retained in heat shield interstices and was vaporizing with enough pressure to impart a rock and roll.

The underside of the orbiter had taken the brunt of the rainstorm. It was thought that a long "soak" of the underside toward the Sun in bottom-to-Sun attitude would bake the moisture out of the tiles and avoid ice formation that would crack them. This procedure accorded with the flight plan, which called for maintaining the orbiter in bottom-to-sun attitude for 33 hours, leaving the payload bay in the shade for that period.

On July 1, when the crew tested the payload doors, one of the doors did not close properly. The same problem, affecting the starboard door, had occurred on STS-3. The crew repeated a maneuver used then to secure the door. They rolled the orbiter so that it would be more evenly heated on all sides. The maneuver worked and the door then closed properly. These experiences showed that the orbiter doors were subject to some distortion by uneven expansion and contraction when the vehicle was kept in one position facing the Sun for long periods.

A report of this mission might not be complete without a note from the ground. The headquarters of the American Philatelic Society at State College, Pennsylvania, received the following letter from a member: "Due to extreme demands on my time, I will no longer be able to serve as chapter representative for the Johnson Space Center Stamp Club. Sincerely, Henry Hartsfield."

Hartsfield was indeed a busy man on STS-4. After the flight, he remarked: "Ken and I believe we've never worked so hard as we did in those seven days."

In addition to monitoring the performance of the ship and experiments, the crew was obliged to keep a food intake record for two high school student science projects approved by NASA. On the night side of the orbit, they searched for thunderstorms to photograph lightning from above the clouds.

For the first time, *Columbia* had been launched due east, so that its orbit was inclined only 28.5° to the equator. The inclination was the lowest of the test series, confining the ground track to 28.5° north and south of the equator. Orbital inclination had been 40.3° on the first mission and 38° on the second and third.

The lower inclination required a change in the first contingency landing site from Europe to Africa—from the Rota Naval Air Station, Spain, to Dakar, in Senegal. It also required a longer cross-range diversion from orbital path to reach Edwards on the 112th orbit. The ground track on this orbit passed approximately 480 miles south of Edwards. Thus, *Columbia* had to swerve northward from the track to approach the landing site, a maneuver well within its 1,100-mile cross-range capability.

Columbia's orbit crossed that of the Soviet manned space station, *Salyut 7*, which was flying an orbit with a higher inclination to the equator. The two vehicles never approached closer than 500 miles, neither was ever visible to the other, and their respective crews made no effort to communicate. These ships that passed in the night were simply watched by North American Air Defense Command and Russian radar.

At 3 A.M. EDT July 1, *Columbia* passed within 8 miles of a Soviet upper-stage rocket, *Interkosmos 14*, which had launched a Cosmos satellite in 1975.

Although space sickness has been epidemic on the shuttle, seeming to be more prevalent in ships large enough to allow crews to move around, the STS-4 crew escaped lightly. Hartsfield complained of a headache and queasiness for a day.

During a rare slack period July 2, Mattingly launched into a discussion with Mission Control about the appearance of the atmosphere. He said he had been noticing how the top of it formed "a very sharp demarcation line" above the Earth's surface. He noticed particularly how completely it blotted out the stars as they sank below it. "It shows how thick the atmosphere is," he said. "It explains why people are so anxious to get their telescopes out here so you don't have to penetrate all that."

During the five American manned spaceflight episodes, Mercury, Gemini, Apollo, Skylab, and the Shuttle, astronauts' comments about the appearance of space and the "sensible" atmosphere have been rare. Frequently, they have described scenes on the ground, but except for storm cloud formations, the appearance of the atmosphere below them is hardly ever mentioned. Although Shuttle test flights were above the part of the atmosphere that can be "felt," *Columbia*'s altitudes were well within the electrified region called the ionosphere, and in that sense the vessel was not entirely out of the atmosphere.

Later on July 2, the mellow mood of the crew was disturbed somewhat by a finding that a nitrogen leak had been detected in one of the tanks serving the cabin atmosphere system. The same leak had been reported during STS-3, and at the same rate, about two pounds an hour.

Shortly after lunch July 3, Mission Control hailed the ship and said: "*Columbia,* be advised you're now on your 100th orbit after going two and a quarter million miles [around the Earth]."

"Yup," responded Mattingly. "Don't even have to change the oil."

During the day, onboard facilities were reviewed. The toilet, which had caused problems on earlier flights, was working. Sleeping arrangements continued to be haphazard, however.

On the first flight, the crew had slept in their seats. Later, as confidence grew in the machine, crewmen stretched out on the deck or in sleeping bags. More permanent "sleep stations" could be set up by fastening sleeping bags to a bulkhead, but Mattingly said he preferred to sleep on the flight deck. Sometimes he crawled into a sleeping bag for warmth when the ship was powered down.

The only problem sleeping on the flight deck, he said, was that "you got to remember to close [shade] the win-

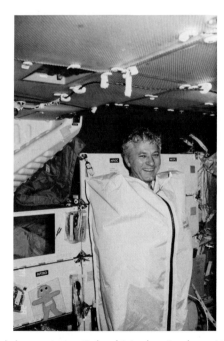

Hartsfield demonstrates *Columbia*'s sleeping bag. Attached to a mid-deck bulkhead, the bag is zipped up the front to keep the astronaut from floating around in the microgravity of orbital free fall. (NASA)

dows. You're tempted to get up every 10 minutes and look out."

Landing day was July 4. In the morning it was announced that President and Mrs. Reagan had arrived at the Dryden Flight Research Center, Edwards Air Force Base, and were en route to the reviewing stand, where they and other notables would watch *Columbia* land.

At that early hour, *Columbia* was already plunging downward toward the atmosphere. Mattingly fired the orbital maneuvering engines at apogee over the Indian Ocean. The effect was to bring perigee down to 0.8 miles from the ground 180 degrees away in California.

As the big space airplane descended silently across the Pacific Ocean, John Young took off from Dryden in a NASA chase airplane to reconnoiter winds aloft and clouds.

Columbia swept past Hawaii and emerged from radio blackout. The crew reported all well, but radar scans showed that the ship was about ten miles off descent trajectory. The flight computers, responding to radio beacons, promptly corrected the path. Both Mattingly and Hartsfield experienced vertigo as air resistance built up and produced the sensation of gravity.

Columbia was coming in rapidly. At 93,000 feet, it was moving at Mach 3, 75 miles to go. Eighty-five thousand feet at Mach 2.6 and 62 miles to go.

Dryden long-range cameras picked up a sea gull at 69,000 feet. It was *Columbia* at Mach 1.6. At 42,000 feet, *Columbia* bounced up and down on the television screens as the long-range cameras tracked it approaching the heading alignment circle. It turned left, descending to 33,000 feet. Air speed was 258 knots. At 28,000 feet, Mission Control spoke: "T. K., you have surface winds at 240 degrees, 12 knots."

T. K. Mattingly: Thank you.

Columbia turned to a heading that aligned it with concrete runway 22. It was the first attempt to land the ship on a concrete runway. Fifteen thousand feet long, it had the same dimensions as the landing strip at Kennedy Space Center. Hitherto, the spaceship had come down on "endless" dry lake and desert runways stretching up to seven miles. Now the landing roll was restricted.

Mission Control advised the crew to keep the ship in the automatic landing mode until it reached 2,500 feet. Then Mattingly would take over. The elaborate autoland system was not fully trusted.

John Young, escorting *Columbia* in the chase plane, called out that the landing gear had come down and locked. "Outstanding!" said he. At that moment, *Columbia*'s main gear touched the concrete with little swirls of dust; then the nose wheel slowly came down until it touched, and the spaceship rolled to a stop.

"Welcome back to Earth," called Mission Control. "That was beautiful."

"That's quite all right," said Mattingly.

It was 12:09 P.M. EDT July 4, 1982. The Space Shuttle Transportation System had completed flight testing. It had taken ten years.

The end of the beginning that July 4 was a historic occasion and obviously deserved to have some words said about it. On hand to say them was President Reagan, who with Nancy Reagan, NASA officials, and astronauts watched *Columbia* come home amid the cheers from 200,000 spectators.

Mr. Reagan compared the final test flight to the "golden spike" that symbolized the completion of the first transcontinental railroad in America a century ago. He referred to the linkup of the Union Pacific and Central Pacific rails, which met at Promontory Summit, Utah, on May 10, 1869, and were ceremonially joined by several spikes of gold and iron.

Then, as the president hailed the new national capability of capitalizing "on the tremendous potential offered by the ultimate frontier of space," a new spaceship appeared in the sky while *Columbia* still exhaled its exhaust gases on runway 22. The new one was Orbiter 099, named *Challenger*. It passed majestically over the bunting-draped reviewing stand and the upturned faces of the president, the First Lady, and their entourage as it rode atop its thundering Boeing 747 transport on the first leg of its airlift from the factory to Florida.

There had been speculation before the flight that the president would use the occasion to announce a space station as NASA's next new project, although White House science adviser, George A. Keyworth II had branded such

speculation as premature. There was no such announcement, but there was no doubt that Mr. Reagan had the space station in mind when he said that it was necessary to look to the future "by demonstrating the potential of the Shuttle and establishing a more permanent presence in space."

Caught between the rock of a mounting budget deficit and the hard place of congressional resistance to cuts in the entitlement programs, the administration must have recognized that the summer of 1982 was not a propitious time to launch a new manned space program that probably would cost as much as the Shuttle. Yet it was clear that despite the success of the Shuttle, the United States had no facility to match the long-duration flight capability of the Soviet space station program. Mr. Reagan appeared to share the view of NASA and Defense Department space station advocates that a continuous U.S. manned presence in Earth orbit was necessary to balance that of the Soviet Union.

During the six-year hiatus in U.S. manned spaceflight from the Apollo-Soyuz Test Project of 1975 to STS-1 in 1981, the Soviets chalked up one man-in-orbit endurance record after another in their Salyut space stations. Later, on December 10, 1982, Lieutenant Colonel Anatoliy Berezovoy and Valentin Lebedev, engineer, returned to Earth after 211 days aboard *Salyut 7*. They broke the 185-day record set in 1980 by Leonid Popov and Valeriy Ryumin in *Salyut 6*. The longest U.S. man-in-space record was 84 days, set by astronauts Gerald P. Carr, Dr. Edward Gibson, and William R. Pogue in *Skylab 4* on February 8, 1974.

A not insignificant aspect of the Soviet endurance records was their demonstration that trained space crews could withstand a microgravity environment long enough to go to Mars without disabling physical impairment. Berezovoy and Lebedev, however, seemed to have reached the upper limit of tolerance after being cooped up seven months in a habitable volume of 100 cubic meters, less than one-third that of *Skylab*. Reports from Russian sources portrayed them as becoming extremely irritable toward the end of the flight. It appears that they were brought down in a hurry, because their landing was less than optimum. Their Soyuz descent capsule

The second orbiter, *Challenger,* en route to Kennedy Space Center aboard the 747 transport July 4, 1982, crosses the Rocky Mountains.   (NASA)

landed at night on a hill during a snowstorm. It rolled down the hill, with the crewmen tumbling head over tincups inside, until it settled in a gully. The storm was so violent that the cosmonauts could not be evacuated by helicopter until morning and spent the night in the back of a truck.

Although it was obvious to administration officials that the United States would have to build a space station in order to match the Soviet manned presence in space, it was equally obvious that this could not be funded while the Shuttle was in development. The Shuttle was taking the lion's share of the NASA budget during this period, even at the expense of new space science initiatives.

I raised this question with Vice President George Bush during a news conference at the Kennedy Space Center in 1981. His response was noncommittal. No decision was in the offing, he said. So it was evident that none would be forthcoming during the gestation period of the Shuttle.

With the successful completion of the test flight program, that constraint was removed. As the Shuttle entered its operational phase with STS,-5, NASA quietly assembled a Space Station Task Force. In mid-1983, its director, John D. Hodge, announced that "the next step for America is to develop a space station that would provide a permanent presence in orbit around the Earth . . .

By 1981, an advanced space station was designed by the Jet Propulsion laboratory with huge solar cell arrays and multiple experimental instruments. It was to be serviced by the Shuttle.   (NASA-JPL)

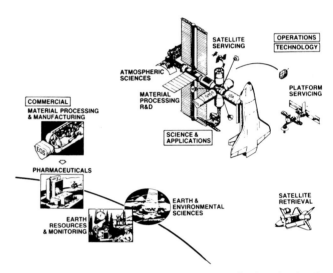

By 1983, a NASA Space Station Task Force had evolved a design for a multipurpose space station serviced by the Shuttle. The station would support scientific observations and experiments, materials processing, and pharmaceutical manufacturing. It would be modular and could be built for crews of 6 or 12 persons.   (NASA)

space would become like the high seas, a medium for manned operations, advancing our technology, adding to our scientific knowledge, and enhancing our security."[2]

If the space station was to be a Go project, it would be under the supervision of Beggs, the new administrator, who was one of the aerospace industry's most experienced executives. Beggs had served during the Nixon administration as NASA associate administrator in the Office of Advanced Research and Technology and later was an Undersecretary of Transportation. He had returned to private industry during the Carter administration, and when President Reagan nominated him for the top post in NASA he was executive vice president and a director of the General Dynamics Corp., St. Louis.

Beggs' background as an engineer and administrator seemed to qualify him exceptionally well to manage a new program that was likely to be as costly and as controversial as the Shuttle program had been. A 1948 graduate of the U.S. Naval Academy, he held a master's degree from the Harvard Graduate School of Business Administration. He had served in the Navy as a lieutenant commander before entering private industry as gen-

eral manager of the Westinghouse Electric Company's Underseas Division in 1955.

One afternoon in July 1983, I talked with Beggs in his seventh-floor office at NASA Headquarters in Washington. He appeared quiet, relaxed, and thoughtful, projecting a firmly decisive image. He spoke with confidence and conviction about his perception of the agency's role and its needs.

On the subject of the space station, he made it clear that this is a firm proposal from NASA and is ranked in the agency's budget cycle. The president, he said, would make the ultimate decision. I inferred from this that the station would be a presidential initiative on the scale of Apollo and the Shuttle.

What was the magnitude of the project? Beggs estimated that in dollars, it would cost as much as the Shuttle had, but considering inflation, it would require only half as much in real resources.* Although Beggs had

*Up to the point of STS-1, Shuttle development had cost $9.91 billion in fiscal year 1982 dollars, which equated to $6.654 billion in fiscal 1971 dollars, according to NASA.

started space station definition studies, the project lacked a commitment from the White House. There was doubt where it ranked on Keyworth's list of priorities.* Beggs recalled that the presidential science adviser had referred to the space station as an instance of the von Braun influence.†

"Wernher always argued that we should have done it long ago," Beggs said. "He was for a Moon base, a Mars landing. Had the nation been willing to commit to spaceflight at the Apollo level this might have been done, but it lost interest in the early part of the 1970s. Now there's interest, but no money. I would be happy with a commitment of $8 to $9 billion a year."

In space program purchasing power, these sums would equate to about one-half of those appropriated during the peak of Apollo development in the mid-1960s. Still, Beggs believed that he could get the space station into development at such a funding level.

Our discussion inevitably shifted to the balance in space activity between the United States and the USSR. Beggs warned that the United States faced increasingly tough competition in space from the Soviets.

"We'll see a Soviet shuttle," he predicted, "and increasingly sophisticated planetary missions."

An example was the launch of *Veneras 15* and *16* in June 1983 by the Soviets to map the cloud-covered surface of Venus by radar. NASA had proposed such a mission two years earlier with a Venus Orbiting Imaging Radar (VOIR) spacecraft, but the project had been shot down by the Office of Management and Budget. The Jet Propulsion Laboratory had then come up with a cut-rate imaging project, the Venus Radar Mapper, but it was not yet funded, while the Soviets were due to put their radar mappers around Venus in October 1983.

*Keyworth was quoted in the July 8, 1983, issue of *Science* as saying he thought the country would take a major thrust in space "very seriously," and "We know we have the technology to build a space station."
†Wernher von Braun, the German-born rocket pioneer who managed the development of the Saturn rocket program, was influential in shaping the Space Task Group report that called for a space station in addition to the Shuttle. He died June 15, 1977.

Beggs discounted comments by some American experts that the Russian radar imaging systems were too crude to achieve impressive results. If all went well, the Russians could steal a march by displaying the first high-resolution photomaps of the Venerian landscape.

"I would bet on their establishing a 10- to 12-man facility in space in this decade—and putting a light on it," he said. "But we don't seem to have the concern about it we had."

The basic problem in funding new NASA projects reflects a 20-year downward trend in funding for research and technology, he said. In the mid-1960s, the nation was spending 2.9 to 3 percent of the Gross National Product on R & D, he noted, but the percentage has dropped although there now seems to be a slight upward trend. "But NASA doesn't have the resources it used to have," he added.

Nationally, our competitive position has deteriorated in terms of mature industry, he said, and in this respect, the space program serves as an asset because it pushes broad-based technologies. Indirectly, he said, it thus causes us to resume a more competitive posture. "I think the indirect benefit is as great as the direct benefits we get," he said.

Beggs cited two main objectives. One is to rejuvenate the space agency, which he characterized as no longer young. The other, of course, is to get the space station off the ground.

"We're middle-aged now, not as vigorous as we were," he said. "We need to upgrade our facilities. The agency is getting old. It needs fresh minds. We're going to be hard pressed to bring in young people by the time the last of the old crowd walks out the door."

With the launch of the space station, Beggs hopes to build a ten-year program for space science and exploration. To implement this, he said, the agency needs a stable budget. There is a lot to be done, said Beggs, and looking back over the years he has known NASA, he added: "We should have done a lot more."

14

We Deliver

Next up was STS-5, the first operational mission. It provided the first opportunity for the Space Shuttle Transportation System, which by this time had cost $10.083 billion in fiscal year 1983 dollars, to demonstrate its purpose as a ground-to-orbit freight line.

Columbia was scheduled to fly the mission in the fall of 1982 while its somewhat lighter-weight twin, *Challenger,* was being prepared in Bay 2 of the Orbiter Processing Facility at the Kennedy Space Center for its first flight early in 1983.

On the first operational mission, the main objective was to deploy two commercial communications satellites from *Columbia* at a cruising altitude of 185 miles. From there each satellite would be boosted by an attached solid-fuel upper stage, the McDonnell-Douglas Payload Assist Module (PAM-D), into an elongated transfer orbit. After the PAM-D was jettisoned, a smaller, solid-fuel "kick stage" would establish the satellite in geosynchronous orbit, 22,300 miles over the equator.

First to be deployed was *SBS 3,* third in a series of powerful communications satellites owned and operated by Satellite Business Systems, Inc. This concern, which provides nationwide residential long-distance telephone service as well as communications network service for big business organizations, was a partnership of the Aetna Life & Casualty Co., Comsat General Corp., and International Business Machines.

The second satellite scheduled for deployment was *Anik C,* Canada's fifth domestic communications satellite and the first of a new and more powerful series. *Anik* is an Inuit (Eskimo) word for "brother," a term reminiscent of the Russian *Sputnik,* "fellow traveler." Earlier Aniks provided communication for widely dispersed settlements in sparsely populated northern Canada, but *Anik C* was to beam television to subscribers' rooftop antennas in southern Canada and the northern tier of the United States, in addition to providing voice, data, and facsimile services.

The Anik system was owned and operated by Telesat Canada, a government-regulated stock company. Both *SBS 3* and *Anik C* were manufactured by the Hughes Aircraft Co., at a cost of $25 million apiece.

Compared to the cost of launching these satellites on an expendable rocket, the Shuttle launch charge was a bargain. In 1981, Satellite Business Systems, Inc., paid $24 million for the launch of its *SBS 2* communications satellite by a Delta 3910 launcher, according to NASA, while the bill for lifting *SBS 3* and its PAM-D upper stage to low orbit on STS-5 was $9 million. A similar charge was assessed for Telesat Canada's *Anik C,* NASA said. The customer provided the upper stage boosters on both Delta and Shuttle launches.

Under the pricing policy set in 1972, the basic charge for the entire cargo bay is $18 million in 1975 dollars until October 1, 1985, when it rises to $38 million in 1975 dollars, or $71 million in 1982 dollars, as computed by NASA. The new price obtains until September 30, 1988, when it is expected to go higher.

The current $18 million charge applies only to civil U.S. and Canadian government users and the European Space Agency. Other users pay a $271,000 surcharge for a reflight guarantee in case the flight fails to deliver the payload to low orbit and a user fee of $4,298,000, a total of $22,569,000. The user fee is charged for government support services and represents a user contribution to the cost of the Shuttle fleet. However, when the price changes in 1985, the user fee will be absorbed in the new price schedule, and government and nongovernment users will be charged equally.

On most flights, the payload bay is shared by two or more users. Each pays a pro rata share plus a small additional fee for unused space needed to separate payloads. The pro rata charge is based on one of two load factors, whichever is greater: the proportion of payload weight to Shuttle lifting capacity (65,000 pounds on a due east launch) or the proportion of payload length to cargo bay length (60 feet). If a 10,000-pound payload is four feet long, the applicable load factor would be weight divided by 65,000 pounds.

The load factor is divided by 0.75 to determine the charge factor, which takes into account the unused space

between payloads. The charge factor is then multiplied by the total payload bay price. Thus, the pro rata share of a 10,000-pound payload where the applicable load factor is weight is:

$$10{,}000 \div 65{,}000 \div 0.75 \times \begin{cases} \$18{,}000{,}000 \text{ (government)} \\ \$22{,}569{,}000 \text{ (nongovernment)} \end{cases}$$

Pricing of the small Getaway Special (GAS) experiment packages, most of them no bigger than a garbage can, is $3,000 for a 60-pound cannister with a volume of no more than 2.5 cubic feet; $5,000 for 100 pounds occupying the same volume; and $10,000 for 200 pounds maximum in a 5-cubic-foot container. These prices are not expected to change in 1985. The Getaway Special program has been fairly successful in attracting universities, small private laboratories, and individuals as Shuttle customers.

As currently in force and as projected, payload charges hardly cover a fraction of actual flight costs. These have been estimated officially at the Kennedy Space Center as an average of $250 million per mission from launch to landing.

The two satellites to be deployed from STS-5 and their attached PAMs added up to 14,600 pounds, a relatively light load. Stowed for launch in *Columbia*'s cargo bay, each was 9 feet 3 inches high, but when established in orbit, with the telescoped aft solar panel extended and the big antenna erected, each measured 21.5 feet high, with a diameter of 7 feet. The extended cylindrical solar panels, sparkling with 14,000 photovoltaic cells, produced one kilowatt of direct current from sunlight. Once established in transfer orbit, with PAM-D jettisoned, *SBS 3* had a mass of 2,462 pounds, including the apogee motor and its fuel. On the same basis, *Anik C*'s mass was 2,557 pounds. *SBS 3* was to be parked in geosynchronous orbit at 94° W. longitude, south of Port Arthur, Texas, and *Anik C* at 112.5° W., south of the Canadian Rockies.

Columbia's first commercial payload was ambitious and significant, a demanding test of the Shuttle's capability of delivering costly and delicate cargo to orbit.

A turntable technique of deploying the satellites was unveiled on STS-5. Instead of being hoisted over the side of the cargo bay by the 50-foot mechanical arm, each satellite and its PAM-D would be thrust into space by powerful springs from a rotating launch pad. Prior to releasing the springs, the turntables would be started revolving at 50 revolutions a minute to provide the spin to stabilize the satellites in orbit.

The method was reminiscent of one used at the dawn of the space age. The upper stages of Missile No. 29 had been set spinning to stabilize them in flight before the Jupiter C roared off Cape Canaveral the night of January 31, 1958 to put *Explorer 1* in orbit.

The First Quartet

STS-5 carried a crew of four, the largest number of astronauts sent into space in one vehicle up to that time. Apollo had carried three, Gemini two, and Mercury one. The Shuttle was designed for as many as seven.

Robert F. Overmyer, pilot of *Columbia*'s first operational mission, STS-5, stands behind his seat on the flight to observe the Earth through the ceiling window as the orbiter crew flies heads down.   (NASA)

The flight commander was Vance D. Brand, 51, a veteran of the Apollo-Soyuz linkup mission in 1975 and the only member of the *Columbia* crew with previous spaceflight experience. Robert F. Overmyer, 46, a colonel in the Marine Corps who had supervised *Columbia*'s heat shield installation for a time, was the pilot.

Two mission specialists were added to the crew. They were William B. Lenoir, 43, a Ph. D. in electrical engineering, and Joseph Allen, 45, a Ph. D. in physics.

Their main tasks were to launch the satellites and take a space walk in the open cargo bay to test NASA's new, million-dollar space suits for shuttle extravehicular activity, or EVA. In the near future, the orbiter would serve as a work platform for space-suited mechanics repairing satellites or performing as orbital steeplejacks in the erection of space structures. The projected EVA on STS-5 was a start in that direction.

After the arm had been removed from the cargo bay and refitting was completed, *Columbia* was rolled out of the Orbiter Processing Facility September 20, 1982, and mated with the solid rocket boosters and the external tank in the Vehicle Assembly Building.

Refitting had been moderately extensive. The number 3 auxiliary power unit and its water spray boiler (which cooled it) were replaced. The number 1 fuel cell battery was replaced by a rebuilt unit. Other items replaced included one of the upward-firing thrusters in the forward reaction control system (it had stuck); all six vernier thrusters; the number 3 inertial measurement unit; and the number 2 Tactical Air Navigation unit. The seemingly unworkable and unfixable toilet was overhauled.

In a perverse way, the standard fixtures accounted for most of the malfunctions in flight. The main engines, which had failed time after time during development, had held up beautifully. Replacements were minimal. The high-pressure oxidizer turbopump on the number 1 engine was replaced as a precaution. It had more running time than any other in the three-engine system. The number 3 engine's high-pressure fuel pump was replaced with a spare. Inspection had indicated it might be subject to ''breakaway torque.''

About 300 tiles were replaced, one-third the number that were replaced after the third mission. Two hundred had been damaged in the hailstorm before the launch of the fourth flight. Another 21 were removed for evaluation and then replaced. Inside the cabin, the explosive devices that hurtled the test pilot–style ejection seats from the flight deck through a roof hatch were removed. That avenue of escape for the pilots was ended in case of explosion at liftoff. The seats and their rails were left in place, however, for later removal at the factory in Palmdale. Two mission specialists' seats were installed, one on the flight deck just behind the pilots' seats and one "downstairs" on mid-deck.

Although satellite deployment was the main objective of the flight, experiments were carried, as on earlier missions. The West German Ministry of Research and Technology had purchased a Getaway Special. It was an experiment in the mixing behavior of two liquid metals, gallium and mercury, in microgravity after heating. The experiment cannister was equipped with a transparent automatic oven and an x-ray machine. Periodic x-ray images showed the effects of microgravity on the dispersion of mercury droplets in liquid gallium. On Earth, at one gravity, the droplets settled to the bottom of the container; in orbit, it was expected that they would remain in the mixture.

Three high school student experiments were carried in NASA's Shuttle Student Involvement Project, a joint venture with the National Science Teachers Association. One, devised by Aaron K. Gillette while a student at Winterhaven (Florida) High School, observed the effect of microgravity on growth of the sponge *Porifera microciona* in seawater containers. A second, prepared by D. Scott Thomas while a student at Richland High School, Johnstown, Pennsylvania, used a TV camera to study the effect of microgravity on surface tension convection of oil in heated containers. The third was an experiment designed by Michelle A. Issel while a student at M. T. Sheehan High School, Wallingford, Connecticut, comparing crystal growth in a solution of triglycine sulphate in microgravity with growth in one gravity. All three experimenters were attending college by the time their experiments were spaceborne, and each was supported by an industrial sponsor.

Up and Away

STS-5 was launched on time, at 7:19 A.M. EST, November 11, 1982, into a 185-mile orbit inclined 28.5° to the equator. The launch was smooth and brilliant, a spectacular sight for more than a half million spectators crowding the beaches, roadways, and lagoon banks around the space center.

A Soviet trawler typical of those that have monitored launches from Cape Canaveral and the Kennedy Space Center is visible on the horizon from the deck of the U.S. Navy's *Range Sentinel,* a patrolling observation ship.   (Rice Sumner Wagner)

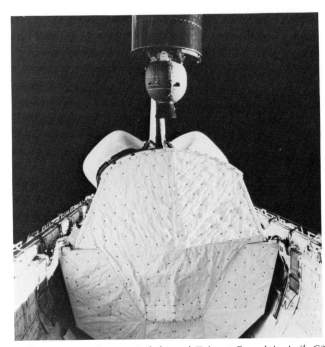

The first satellites deployed into space by the Shuttle system are *Satellite Business System 3* (left) and Telesat Canada's *Anik C3* (right) from *Columbia* on the STS-5 mission. These photographs were taken as the Satellites were ejected from their protective shrouds in *Columbia*'s cargo bay during the STS-5 mission.

(NASA)

On the sixth orbit, eight hours after liftoff, Lenoir released *SBS 3* and its PAM-D booster from the revolving turntable. Onboard television showed the payload leaving the cargo bay like a slow-moving, spinning projectile, at three feet a second, over the cloud-decked Atlantic Ocean.

Overmyer had maneuvered *Columbia* so that its starboard wing pointed Earthward and its open payload bay faced opposite the direction of flight. Ship and satellite rapidly separated, although the satellite continued following *Columbia*'s ground track. Fifteen minutes after deployment, a burn of *Columbia*'s orbital maneuvering engines lifted the ship's apogee to 201.4 miles. The ship was rolled so that its heavily insulated underside was toward the satellite. The maneuver protected the cabin windows and cargo bay from PAM-D exhaust when the rocket motor fired automatically 45 minutes after deployment. By that time, *Columbia* had moved 16 miles above and 17 miles ahead of *SBS 3*. Later, after PAM-D

was jettisoned, engineers at COMSAT Launch Control Center, Washington, D.C., signalled the apogee motor, or kick stage, to fire the fifth time transfer orbit apogee was reached. The additional energy imparted to the satellite at apogee raised its perigee to circularize the orbit at geosynchronous altitude. *SBS 3* was in business.

The next day, on *Columbia*'s 22d orbit, Allen deployed *Anik C* over the equatorial Pacific Ocean. Away it went, spinning like a top. Later, the apogee motor was fired by radio signal from Ottawa. *Anik C* was in business.

Both deployments were so neatly done that they looked like a demonstration contrived in a movie studio, but they were real enough to elicit congratulations from Mission Control. In response, the crew displayed a sign to the TV camera. It said:

ACE MOVING CO.
WE DELIVER

Their satellite deployment mission accomplished, the crew displays its famous "We Deliver" message. Holding the sign is flight commander Vance D. Brand. Floating around him clockwise are William B. Lenoir, mission specialist; Overmyer; and Joseph P. Allen, specialist.

(NASA)

The Space Shuttle Transportation System was in business.

On the fourth flight day, Allen and Lenoir got ready to test the new space suits. Lenoir had suffered a bout of space sickness, medically termed "space adaptation syndrome," which has plagued some astronauts and cosmonauts but not others since the beginning of manned spaceflight. Gherman Titov was its first victim on the 17-orbit flight of *Vostok 2* in August 1961. However, by flight day four, Lenoir had recovered and was ready for the extravehicular activity.

In the Murphy's Law tradition of spaceflight, the exercise was halted by minor breakdowns. As Allen struggled into his space suit, the fan motor malfunctioned. It would run slowly, surge, and shut down. Allen's EVA was scrubbed. Lenoir also experienced suit trouble. The oxygen regulator failed to supply adequate pressure. His EVA was scrubbed, too. Mission Control said that the exercise was deferred until the next flight.

Subsequent investigation, according to NASA, traced the fan motor glitch in Allen's suit to a malfunctioning sensor in the motor electronics. The oxygen regulator problem in Lenoir's suit was traced to the failure to install two small plastic locking devices during assembly. Without them, a threaded ring that holds a spring in place had backed off its proper adjustment. Henceforth, the

space agency vowed, space suits would be tested before scheduling EVAs.

Otherwise, the flight of STS-5 was free of problems. Five days, 2 hours, 14 minutes, and 25 seconds after liftoff in Florida, *Columbia* was landed by its pilots on concrete runway 22 at Edwards Air Force Base, California. The time was 9:33 A.M. EST, November 16, 1982. A partial test of the automatic landing system was carried out during descent, but the final approach was flown manually.

At touchdown, Brand called capsule communicator Roy Bridges in Houston.

SC: Hey, Roy. Are we down? That was a . . . are we on the ground?
CC: Absolutely. It was beautiful and you certainly lived up to your motto this flight. Welcome home!
SC: Yessir. We delivered!

Epilogue

The climactic voyage of *Columbia* was its ten-day flight November 28–December 8, 1983, with Spacelab, the European Space Agency's contribution to the Shuttle, in the cargo bay. Spacelab consisted of the laboratory module, 23 feet long and 13.3 feet in diameter, and the 9.5-foot-long open pallet on which was mounted an array of instruments.

The mission, STS-9–Spacelab 1, was the longest and most complex of the Shuttle era thus far. Its six-man crew was the largest that had been launched into space in one vehicle up to that time. It carried exerpiments designed by more than 100 scientists and institutions in Austria, Belgium, Denmark, France, Germany, Italy, Japan, the Netherlands, Spain, Sweden, Switzerland, the United Kingdom, and the United States.

Although an unprecedented computer failure delayed the landing for 7 hours and 49 minutes, *Columbia*'s performance on this, its sixth flight, was hailed by the Shuttle directorate as proof of the durability of the orbiter design. "I really like used spacecraft," remarked Lieutenant General James A. Abrahamson, NASA Associate Administrator for Space Flight, after John Young, the flight commander, landed *Columbia* smoothly on Rogers Dry Lake bed at Edwards Air Force Base.*

The computer problem at first looked ominous. As the crew was preparing to land on the morning of the tenth flight day, two of the five on-board IBM computers that control guidance, navigation, and other orbiter functions shut off. This seemed to have occurred at the moment the ship was jarred by an unusually powerful firing of an attitude control thruster.

Shortly thereafter, one of three inertial measuring units that feed flight data to the computers also failed. The landing was waved off by Mission Control while computer experts tried to find out what was happening.

Young was able to bring one of the two failed com-

*Abrahamson had been promoted from major general to lieutenant general July 4, 1982.

puters back on line, but the other remained dead. After Mission Control determined that four of the five computers in the data processing system were available for reentry and landing, the crew was given a "Go" for reentry later in the day.

Redundancy in the system is such that *Columbia* could have returned safely with one of the five computers operating. Nevertheless, the Shuttle directorate was shaken by the incident, especially when the computer that Young had managed to restart failed again as *Columbia*'s wheels touched down on lake-bed runway 17.

Although computer glitches had occurred at launch and on earlier Shuttle flights, this was the first complete computer breakdown in an orbiter during spaceflight.* The problem followed an earlier series of malfunctions and mishaps in the Shuttle flight program after STS-5 that had delayed the Spacelab flight three months.

While *Columbia* was being refitted with new and more powerful engines and adapted for Spacelab and a larger crew, the second orbiter, *Challenger,* flew the sixth, seventh, and eighth space missions. Although successful, they were beset by breakdowns. In actuality, each of these missions represented an advance in the complexity of shuttle operations, and the problems that plagued them could charitably be considered the price of a process of rapidly developing the Shuttle's potential.

The first of three tracking and data relay satellites (TDRS) is shown (rear) with two others in earlier stages of construction at the TRW, Inc., factory, Redondo Beach, California. (TRW)

STS-6: *Challenger*

A prime requirement of the Spacelab mission was the establishment of an advanced space-to-ground communications facility—the tracking and data relay satellite system, called the TDRSS, or "tee-dress," for short. The system was to be composed of two powerful communications satellites 130° apart and a spare between them in geostationary orbit. The satellites would be linked to the NASA communications network through a ground station at White Sands, New Mexico.†

TDRSS had been in development for nearly a decade to replace most of the 15 ground stations in NASA's 20-year-old space tracking and data network (STDN) and to expand the coverage and speed of communications, especially data transfer, between orbiting spacecraft and the ground. For Shuttle flights, the new system had dra-

---

*The two computers that failed on the last flight day and the two leaking auxiliary power units that caused a fire to smolder in an aft compartment when *Columbia* landed were removed and sent back to their manufacturers. Unnoticed by the crew, evidence of the fire was not discovered until engineers boarded *Columbia* at Dryden Flight Research Center for postlanding inspection. Failures of computers 1 and 2 in the set of five that control *Columbia* resulted from tiny slivers of gold and solder that broke loose in the hardware and short-circuited electronic components, according to NASA. The diagnosis was made by the manufacturer, International Business Machines. The power unit leaks were traced to a deformed "O" ring in the fuel control valve, NASA reported.

†The satellites were manufactured by the TRW Defense and Space Systems Group, Redondo Beach, California, and the antenna systems by the Harris Corporation, Government Communications Division, Melbourne, Florida.

Challenger's cargo bay carried the first tracking and data relay satellite, *TRDS A,* on STS-6. It is shown being inspected in the Kennedy Space Center Operations and Checkout Building.
(Rice Sumner Wagner)

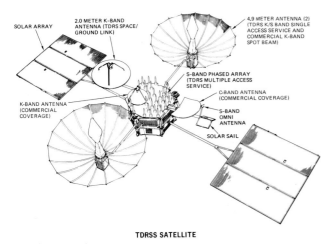

Diagram of the tracking and data relay satellite. Two in geostationary orbit and one spare will replace most of NASA's ground stations. (TRW)

TRDS A is mated to the inertial upper stage booster at Kennedy Space Center's vertical processing facility. (Harris)

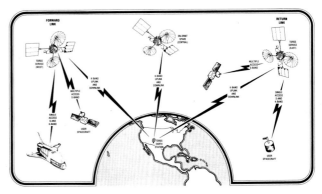

Diagram showing how the tracking and data relay satellite system will function as a satellite and Shuttle link to Earth. (TRW)

Silhouetted against the New Mexico sky, three Earth station antennas from part of the tracking and data relay satellite system control center at White Sands, New Mexico.　　　(Harris)

Challenger moves toward the launch pad in a dense fog for the STS-6 mission.　　　(NASA)

matic advantages. It provided communication along 75 percent of the orbit, compared with the 15 percent provided by the old ground network. It was designed specifically to relay high-speed scientific data (at a rate up to 48 megabits a second) from Spacelab's computerized data transmission system and also from such advanced satellites as *Landsat 4.* Ultimately, TDRSS would provide linkage with NASA's projected space station.

Deploying the first of these 5,000-pound satellites, *TDRS A,* was the task of *Challenger's* inaugural mission, which was scheduled for launch January 30, 1983. But engine problems reminiscent of those that had delayed the initial launch of *Columbia* appeared in *Challenger* and delayed its launch 74 days. First, a leak of gaseous hydrogen was found in the main combustion chamber coolant outlet manifold of engine 1 after a 20-second firing test while *Challenger* was on the launch pad December 18, 1982. A replacement engine shipped to KSC from the National Space Technology Laboratory in Mississippi developed a leak in the liquid oxygen heat exchanger.

While KSC and contractor engineers awaited a second replacement engine, hydrogen leaks were detected in the fuel lines of engines 2 and 3 where metal sleeves had been brazed (as an afterthought) to prevent chafing during launch. In each line, the installation of the sleeve, supposedly an improvement, had caused a hairline crack. It was a pointed instance of an international maxim in space work: "The better is the enemy of the good."

Engines 2 and 3 were repaired in the Orbiter Processing Facility by removing the sleeves and replacing the cracked fuel lines. When the second replacement for engine 1 arrived, its fuel line sleeve was removed, although the fuel line did not appear to have a crack.

While *Challenger* stood engineless on the pad, a windstorm February 28, 1983, breached a seal between the open cargo bay and the payload changeout room of the rotating service structure in which the orbiter was cradled. *TDRS A* was contaminated by sand and dust and had to be cleaned. After agonizing conferences, it was decided that this could be done in the adjacent changeout room without removing the satellite from the launch pad.

At last, the countdown began at the end of March and *Challenger* was launched at 1:30 P.M. EST April 4, 1983, into a 178-mile-high orbit inclined 28.45° to the equator. It carried a cargo with a mass of 46,615 pounds, the largest thus far. The major payload element was *TDRS A* and its inertial upper stage (IUS) booster, a sophisticated rocket manufactured for the Air Force by the Boeing Aerospace Corporation of Seattle.

Challenger performed without any major flaw on its inaugural, five-day flight, but the upper stage rocket, which was being used by NASA for the first time, did not. Ten hours after the launch, the satellite and its booster were deployed by the flight's mission specialists, Story Musgrave, 47, a physician, and Donald K. Peterson, 49, a retired Air Force colonel. The flight commander, Paul J. Weitz, 50, a retired Navy captain and veteran of the first manned *Skylab* mission, and Air Force Colonel Karol J. Bobko, 45, pilot, moved *Challenger* away from the payload. The combined mass of the satellite and IUS was 37,000 pounds.

Fifty-five minutes later, the IUS first stage fired its solid propellant motor for 2 minutes, 33 seconds, lifting the satellite into a highly elliptical transfer orbit. At 6 hours and 12 minutes after deployment, the IUS second stage began to fire to complete the transfer to geostationary orbit, but Murphy's Law intervened. At 70 seconds into the burn, which was programmed to last 104 seconds, communication with *TDRS A* was lost.

After several hours of searching, controllers at the Goddard Space Flight Center in Maryland and the TDRSS ground station in New Mexico began to receive intermittent signals that showed that the spacecraft and IUS second stage were tumbling. The owner and operator of the satellite, Space Communications Company, better known as SPACECOM, faced a half-billion-dollar loss.*

By April 5, controllers had reestablished control of the satellite. They commanded it to separate from the IUS second stage and extend antennas and electricity-generating solar panels. Tracking data showed that the IUS misfire had left the satellite in an 18-hour orbit, with a perigee of 13,574 miles and an apogee of 21,970 miles. The orbit oscillated 2.4° north and south of the equator. In order to function adequately, *TDRS A* had to be moved into a 24-hour orbit precisely over the equator, where its revolution would be synchronous with the Earth's rotation.

Without a booster, however, moving the satellite into geostationary orbit posed a unique problem. Inspired by desperation, NASA and contractor engineers commenced calculations for a unique solution. It called for a novel feat of boosting the 2.5-ton satellite 8,662 miles from its transfer orbit perigee, to 22,236 miles, and circularizing the orbit at that altitude by firing the satellite's reaction control thrusters in carefully planned sequences. If this could be done, it would establish the satellite in an operational geostationary orbit and make it possible to fly Spacelab sometime in 1983.

TDRS A carried 24 of these small reaction engines, each producing about one pound of thrust through a thimble-sized nozzle. Burning them for a major altitude change was a long shot, but it was the only play left in this high-stakes game.

The reaction control system used hydrazine fuel, and there was 1,300 pounds of it, more than enough to con-

STS-6 crew, left to right: Donald H. Peterson, mission specialist; Paul J. Weitz, commander; Dr. Story F. Musgrave, mission specialist; and Karol J. Bobko, pilot. (NASA)

*SPACECOM was a partnership affiliated with Continental Telecom, Inc., Fairchild Industries, Inc., and Western Union Corporation. It had purchased *TDRS-A*, valued at $416 million, from TRW and the IUS, valued at $77 million, from Boeing. SPACECOM has a ten-year contract with NASA leasing TDRSS service for $250 million a year.

trol the spacecraft for its projected 10-year life in orbit. There would be enough fuel for the boost and, hopefully, enough left for future control.

After weeks of observation and calculation, controllers began 58 days of intermittent maneuvers. Slowly, *TDRS A*'s orbit rose; gradually, its elliptical shape became more and more circular. On July 1, 1983, the team at White Sands initiated the final burn that lifted perigee the final 23 miles to synchronous altitude. Apogee, which also had risen during maneuvers, was corrected downward 2 miles. The satellite's drift across the equator was stopped. It reached 24-hour synchronous orbit at 67° west longitude, over Brazil. From there, it was a small matter to shift it to 41° west longitude, its Atlantic station. *TDRS A* became designated now as *TDRS 1* and Spacelab became a Go mission for 1983—or so it appeared.

NASA had planned to deploy *TDRS B* on STS-8, *Challenger*'s third mission, in the summer of 1983 so that a second satellite at 171° west longitude over the Pacific would be on station to handle Spacelab data traffic in the fall, but *TDRS B* was grounded pending the investigation of the IUS failure. This was dragging on well into 1984.

ESA and NASA agreed that it was feasible to fly Spacelab with only a single TDRS, using the ground station network to a greater extent than had been planned. However, a long testing process for *TDRS 1* made it necessary for NASA to postpone the STS 9–Spacelab 1 launch from September 30 to October 28, 1983.

Challenger had moved smoothly through the STS-6 mission, landing at Edwards April 9. Musgrave and Peterson performed an extravehicular sortie into the open cargo bay April 7. This time, their space suits worked perfectly, they reported, during a 3-hour, 45-minute space walk. They rehearsed tasks that would be required to repair the *Solar Maximum Mission (SMM)* satellite on a flight scheduled for April 1984.*

*The *SMM* satellite, designed to monitor solar activity at the height of the sunspot cycle, failed ten months after it was launched from Cape Canaveral February 14, 1980. Its data had shown slight variations in the solar constant, the totality of sunshine reaching the top of the atmosphere. Further investigation of this phenomenon had been planned on Spacelab.

A highlight of the STS-6 mission was the first extravehicular activity aboard a Shuttle. Musgrave (left) and Peterson tested handholds and rehearsed repair tasks in the open cargo bay. Mexico is passing below. (NASA)

Experiments in electrophoresis and production of the monodisperse latex microspheres and the optical survey of lightning were continued. Three Getaway Special payloads tested microgravity effects on the formation of snow crystals, a project sponsored by a Tokyo newspaper; on plant seeds, a project sponsored by the Park Seed Company of Greenwood, South Carolina, and on the growth of microorganisms and metal processing, in an array of experiments devised by U.S. Air Force Academy cadets.

STS-7

In contrast to the 274 days it had taken to prepare and repair *Challenger* at KSC for its first flight, the spaceship was processed, mated with tank and boosters, and launched on its second flight in 64 days. Liftoff was on time at 7:33 A.M. EDT June 18, 1983, but the landing did not follow the flight plan. *Challenger* had been scheduled to be the first orbiter to land at the Kennedy Space Center, but fog and drizzle at the landing site on the morning of June 24, when the landing was scheduled, forced Mission Control to wave the crew off as they prepared for reentry and circle the Earth again to come down at Edwards.

The focus of public interest on the six-day mission was America's first woman astronaut to fly in space, 32-year-old Sally K. Ride. The articulate brunette physicist was one of three mission specialists in the crew of five. Crippen, the pilot on *Columbia*'s first voyage, was in command. Frederick H. Hauck, 42, a Navy captain, was the pilot. Ride's fellow mission specialists were John M. Fabian, 44, an Air Force colonel, and Norman E. Thagard, 39, a physician. Dr. Thagard was added to the crew after

STS-7 carried the first five-member crew: First row, left to right, are Ride, mission specialist; Robert L. Crippen, commander, and Frederick H. Hauck, pilot. Standing are John M. Fabian (left) and Dr. Norman E. Thagard, mission specialists. (NASA)

it had been selected, to study space sickness, which seemed to be endemic on Shuttle flights.

As the first American woman in space, Ride was exposed to the gamut of media exploitation, including a syndicated newspaper biography that depicted her as America's girl next door. None of this attention ruffled her composure or strained her quick good humor. She projected an image of quiet, steady competence, and judging from the public's response they loved it.

As *Challenger* reached orbit, the Mission Control communicator asked Ride how it was. "Have you ever been to Disneyland? Well, this is definitely an 'E' ticket!" she replied. (An "E" ticket is good for all the attractions.)

From the Soviet Union came a telegram addressed to Ride and signed by Valentina Tereshkova, the first woman to fly in orbit, 20 years earlier. She had been a 25-year-old textile worker and parachutist who had answered a call for female cosmonaut trainees. Tereshkova had flown 48 orbits in the single-seater *Vostok* 6 spacecraft June 16–19, 1963. Now 45 and a mother, she congratulated Ride, saying: "It gives me pleasure to know that a third representative of this planet's women, now from the United States of America, is in outer space today."* The

Astronaut Sally K. Ride was the first American woman to fly in space. She was a mission specialist aboard *Challenger*'s second flight, STS-7.   (NASA)

*New York Times, June 23, 1983, p. 13.

second woman in space was cosmonaut Svetlana Sav-itzkaya, a 34-year-old test pilot, who visited the *Salyut 7* space station with cosmonauts Leonid Popov and Alek-sander Serebrov in the *Soyuz T-7* spacecraft in August 1982.

Ride and Fabian deployed Telesat Canada's *Anik C* and Indonesia's *Palapa B* communications satellites, which were then boosted to geostationary orbit positions by their payload assist module (PAM-D) upper stage boosters. Ride and Fabian successfully operated the re-mote manipulator arm to deploy and retreive a 2.5-ton West German instrument platform called SPAS (Shuttle Pallet Satellite). It was a prototype of future free-flying

The first picture of an orbiter in flight was this one of *Chal-lenger* on STS-7, taken June 22, 1983, from the Shuttle Pallet Satellite, a West German experimental platform, which car-ried a 70mm automatic camera. (NASA)

Hughes Aircraft Co. technicians adjust the spin mechanism of the spin-stabilized section of the *Palapa B* communications satellite. The machine was deployed in orbit from *Challenger* (STS-7) for Indonesia to provide voice, video, telephone, and data services for Indonesia, the Philippines, Thailand, Malay-sia, and Singapore. (NASA)

platforms that would carry telescopes, radiometers, and other instruments and would be serviced periodically by Shuttle flights.

Challenger carried experiments in semiconductor crystal growth, alloys formed only in microgravity, and the processing of glass-forming melts by acoustic levita-tion, in which the sample was suspended by sound waves. These experiments were packaged in a materials Exper-iment Assembly unit in the cargo bay, which operated automatically. They were sponsored by NASA's Office of Space Science and Applications and a West German industrial group. There were also seven Getaway Special payloads containing 22 experiments. These ranged from tests of cosmic ray effects on plants to the behavior of a colony of carpenter ants. The ants perished. In the mid-deck were the electrophoresis machine and the latex re-actor on another round of tests.

STS-8

Without the second TDRS, *Challenger*'s third flight carried a reduced cargo of 26,163 pounds, the lightest of its three missions. It was launched in rain-swept darkness at 2:32 A.M. EDT August 30 from Kennedy Space Center and landed in darkness at 3:44 A.M. EDT Septem-ber 4 at Edwards. High-intensity xenon arc lamps illuminated the runway. Without lights of its own, *Challenger* slipped silently like a shadow into final approach, visible only to the infrared cameras, its nose, still hot from reentry, glowing brightly on the screen. It was the Shuttle's first night launch and landing.

Truly, who had been STS-2's pilot, commanded the flight, and Daniel C. Brandenstein, 40, a Navy commander, was the pilot. The mission specialists were Dale A. Gardner, 34, a Navy lieutenant commander; Guion S. Bluford, Jr., 40, an Air Force lieutenant colonel; and William E. Thornton, 54, a physician. Dr. Thornton was the oldest astronaut and Bluford, a Ph.D. in aerospace engineering and distinguished alumnus of Pennsylvania State University, the first black astronaut to fly in space.

Challenger flew a 172.5-mile-high orbit inclined 28.45° from the equator. The early morning launch was required for optimum results in launching the *INSAT B* communications satellite and its PAM-D booster for the government of India. The satellite was established in geostationary orbit at 74° east longitude.

One of the crew's principal tasks was to test *TDRS 1*

The first night launch of the Shuttle illuminated the sky for miles August 30, 1983, despite clouds and drizzle. It was *Challenger*'s third mission, STS-8.   (NASA)

Fisherfolk trying their luck in the shallows of Port Canaveral casually observe the "big one" as it passes. It is the casing of one of the solid rocket boosters that helped launch *Challenger* on STS-8. The spent rocket is being towed to the processing facility after being fished out of the Atlantic about 160 miles off shore.   (Robert McDonald, *Today*)

A five-member crew also flew STS-8. From left to right are Daniel C. Brandenstein, pilot; Dale A. Gardner, mission specialist; Richard H. Truly, commander; and Dr. William E. Thornton and Guion S. Bluford, Jr., mission specialists. Bluford was the first black astronaut to fly in space.   (NASA)

in all of its communication modes. Once more the remote manipulator arm was tested, this time raising and swinging around a dumbbell-shaped, 8,500-pound test device. It was planned to use the arm to retrieve the *SMM* satellite in 1984 and redeploy it after repair.

Experiments testing the effects of exposing spacecraft materials to molecular and atomic oxygen bombardment at Shuttle cruising altitudes and testing a new technology for satellite temperature control were carried in the cargo bay. There were four Getaway Special payloads, experimenting with cosmic ray effects on integrated circuits and on exposure of ultraviolet-sensitive film, and with the effect of microgravity on crystal growth and on artificial snow formation, a repetition of the Japanese experiment on STS-6.

Dr. Thornton observed himself and his fellow crewmen for symptoms of space sickness. He also monitored a cageful of rats, the subject of a microgravity behavior experiment devised by a high school student in Hyde Park, New York, and sponsored by the Air Force School of Aerospace Medicine. It was altogether an uneventful flight.

STS-9—Spacelab 1

Communications tests with *TDRS 1* showed that, while one of its electronic systems had failed, the satellite could support the Spacelab mission. After nearly a year in preparation, *Columbia* was rolled out to the launch pad September 28, 1983, for the October 28 launch. A new problem came to the surface when the STS-8 solid rocket booster casings were fished out of the ocean and returned to KSC for inspection and refurbishment. Unexpectedly severe erosion was noticed in the phenolic carbon cloth insulation of the nozzle of the left-hand booster. Layers of this cloth 3 inches thick and backed by 0.05 inches of phenolic glass line the nozzle of every solid rocket booster to prevent its aluminum structure from being melted by the booster's flaming exhaust. During the 125 seconds the motors are firing, the exhaust reaches 5,600° F.

Normally, about one-half of the laminated cloth erodes and chars, carrying the heat away from the metal. Inspection of the STS-8 left-hand booster nozzle showed that a hole had been burned into the insulation to within 0.2 inches of the metal at the nozzle inlet. Questioned at a prelaunch news conference November 26, Larry B. Mulloy, a booster engineer of the Marshall Space Flight Center, estimated that the lining would have burned all the way through in another 16 to 17 seconds and then, he said, the exposed metal would have sublimated instantly. The insulation had ablated (flaked off) at twice the expected rate, he said. Erosion of the right-hand booster nozzle insulation was normal.

What would have happened if the left-hand nozzle insulation had failed during the firing is speculative. Media reports suggested that had the nozzle been destroyed *Challenger* might have veered off course or, if flaming exhaust had reached the external tank, exploded. General Abrahamson expressed a conservative view: "It was not that we were just seconds away from disaster," he told the news conference. "Frankly, it's kind of problematical about what could have happened. Our major problem was understanding this phenomenon."

Mulloy said that he doubted that steering control would

have been affected. If a burn-through had happened after 121 seconds, the crew would not have noticed the difference, he said. He ruled out the likelihood of an explosion.

Nevertheless, it was apparent that flight safety margins were seriously compromised by the anomaly, and the possibility of a repetition of it on STS-9 would not be tolerated. Investigation by engineers for NASA and the contractor, the Morton-Thiokol Company's Wasatch Division, found that the left-hand booster nozzle on STS-8 had been lined with material fabricated by a different supplier using a slightly different process from that used in fabricating the right-hand nozzle insulation. What puzzled investigators was the fact that both processes met specifications, according to Mulloy, yet the one used for the left nozzle insulation resulted in material with a slightly higher volatile component. No one seemed to know why.

Scale-model rocket firing tests of the material from the same batch as that used to line the left nozzle yielded ambiguous results except in one test, where the abnormal erosion rate was duplicated. That was enough to cast doubt on the integrity of the batch. When it was then discovered that the STS-9 right-hand booster nozzle had been lined with insulation from the suspect batch, the October 28 launch date went up in smoke. The Shuttle directorate ruled that the nozzle had to be replaced by one lined with insulation from a batch that had proved satisfactory. In order to do this, STS-9 had to be taken down from the launch pad, hauled back to the Vehicle Assembly Building, and disassembled so that the right booster's 11-ton aft assembly could be replaced. *Columbia* was rolled back into the Orbiter Processing Facility for storage during this process.

Meanwhile, a traffic jam in orbiters developed at KSC. The third orbiter, *Discovery*, was due to arrive from California in early fall, but now there was no room for it in the OPF while *Columbia* roosted in one of the two bays and *Challenger* was being serviced in the other.

By the end of October, the nozzle exchange had been completed. *Columbia* was moved back to the VAB and mated to its boosters and external tank. On November

8, STS-9 was once more moved to the launch pad. The next day, *Discovery* arrived at KSC aboard its Boeing 747 ferry and was wheeled into the berth *Columbia* had vacated.

STS-9 was launched at 11 A.M. EST November 28, 1983, with the heaviest returnable cargo in the history of spaceflight. The crew was composed of four astronauts and two career scientists, who were designated as payload specialists. One was the first European to go into space aboard an American vessel, Ulf Merbold, 42, a West German physicist. The other was an American biomedical engineer, Byron K. Lichtenberg, 35.

In addition to John Young, who had commanded *Columbia*'s first flight, the astronaut crew consisted of Air Force Major Brewster A. Shaw, Jr., 38, pilot, and two mission specialists, Robert A. R. Parker, 47, an astronomer, and Owen K. Garriott, 53, an electrical engineer and veteran of *Skylab 3*. The crew was divided into two 12-hour shifts with Young, Parker, and Merbold on the first, or "red," shift and Shaw, Garriott, and Lichtenberg on the second, or "blue," shift.

In addition to *Columbia*'s new engines, uprated from 100 to 104 percent of the original design thrust, an airlock and tunnel adapter had been installed in the payload bay forward bulkhead to enaʋe the crewmen to

Discovery, the third orbiter, is parked in the Vehicle Assembly Building to allow Spacelab to be unloaded from *Columbia* in the Orbiter Processing Facility after the spaceship returned to Kennedy Space Center. (NASA)

Spacelab is launched aboard *Columbia* at 11 A.M. EST November 28, 1983. (NASA)

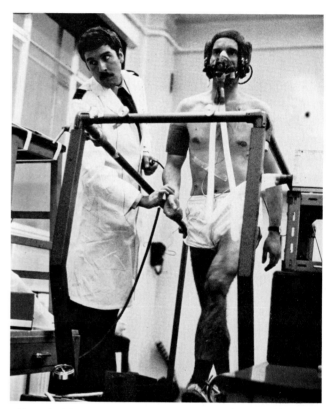

Ulf Merbold, a West German physicist, is the first European to fly in an American space vehicle. He served as payload specialist on the first Spacelab mission, STS-9. As part of the European Space Agency's selection process, he undergoes an exercise stress test at the Royal Aircraft Establishment, Farnborough, England. Merbold was one of three ESA Spacelab candidates chosen for training in 1978 from a field of 2,000 applicants. (I. Berry/Magnum for ESA)

transfer from the cabin to the Spacelab module through an 18.8-foot pressurized tunnel 40 inches in diameter. On STS-5, *Columbia*'s previous flight, three crewmen had been seated on the flight deck and one on the mid-deck. Two more seats were added on mid-deck, as were three bunk beds, three hammock-type sleeping bags, lockers, a small galley and pantry, and a mini-washstand for freshening up and brushing teeth. A single toilet served the group, but no complaints about it were heard in the voluminous air-to-ground conversation that *TDRS 1* made possible.

Spacelab and its support equipment comprised a payload bay mass of 33,252 pounds. Although it was not the heaviest lifted by the Shuttle to orbit, it was the largest mass yet brought back from orbit and weighed a half-ton more than the maximum returnable weight of 32,000 pounds calculated in 1981.*

Columbia was 7.5 tons heavier at landing than it or *Challenger* had been on any previous missions, according to NASA.

Because of its high landing weight, *Columbia* was scheduled to land on the seven-mile-long runway 17 on Rogers Dry Lake bed. It would have plenty of room to roll in the event its landing speed was high, but this precaution turned out to be unnecessary.

Columbia flew in a circular 155-mile orbit inclined 57° to the equator, the highest inclination of any American manned spaceflight. The ground track extended from the subarctic to the subantarctic. It passed through Scotland, Denmark, the Soviet Union, Sakhalin Island, Kamchatka, the Aleutians, and southern Alaska, regions never

Four payload specialists chosen to participate in the first Spacelab mission flown by *Columbia* (STS-9) are (left to right) Merbold; Byron Lichtenberg, U.S.; Wubbo Ockels, Netherlands; and Michael Lampton, U.S. Merbold and Lichtenberg monitored and frequently corrected malfunctions in the Spacelab experiments aboard *Columbia* while Ockels and Lampton monitored experiments from the ground at Houston. They are shown in a training mock-up of Spacelab. (NASA)

before visible to a Shuttle crew. In the southern hemisphere, the track reached as far south as the northernmost extension of the antarctic ice pack and the stormy region of the antarctic convergence, where the Atlantic, Pacific, and Indian Oceans become the cold Antarctic Ocean.

Spacelab was the most sophisticated scientific vehicle flown by NASA since *Skylab*. Its 38 experiment facilities provided instrumentation for 71 scientific investigations. ESA sponsored 58 of them, 35 in materials science, and NASA sponsored 13.

The emphasis on materials science by the ESA countries reflected Western Europe's interest in the economic potential of space processing technology, a strong motive for the billion-dollar European investment in Spacelab. German research firms and institutions devised 13 materials science experiments; French ones, 8. The balance was provided by research centers in Austria, Belgium, Denmark, Italy, the Netherlands, Spain, Sweden, and the United Kingdom.

Equipment for most of the processing experiments with alloys, glass, metallic emulsions, and crystal growth was contained in an elaborate Materials Science Double Rack Facility furnished by ESA for the module. The facility provided three furnaces and a device for experiments in fluid physics.

The array of scientific instruments on the U-shaped pallet adjoining the aft end of the laboratory module included the far ultraviolet space telescope (FAUST) for star observation, a gas scintillation x-ray detector, phyrheliometers measuring the intensity of solar radiation, particle accelerators emitting high-intensity electron and ion beams and streams of netural gas for plasma investigations, magnetic field sensors, a cosmic ray detector of packaged plastic sheets, an imaging spectrometer, a 16mm camera to photograph clouds, a radar imaging antenna, and a television camera and photometer to photograph faint light emissions from the atmosphere. Three pyrheliometers were contained in the active cavity radiometer mentioned in chapter 4 to measure total solar radiation reaching the top of the atmosphere. French and Belgian instruments designed to observe this radiation (the solar constant) and detect any variations in its intensity were also mounted on the pallet.

The payload specialists operated a Very Wide Field Camera (supplied by ESA) mounted in the module's sci-

Wrapped in thermal blankets, Spacelab is settled in *Columbia*'s cargo bay. The open instrument pallet extends aft.

(NASA)

Buttoned up in their protective shrouds in *Columbia*'s cargo bay are the Spacelab module and the tunnel connecting it to the orbiter cabin.   (NASA)

entific airlock to photograph the sky in ultraviolet light. ESA also provided a high-resolution Metric Camera designed for photomapping of the ground from orbit.

Five experiments in plasma physics probed the ionosphere, the region of electrically charged particles through which *Columbia* was flying. Two accelerators on the pallet fired high-intensity beams of electrons and ions into the ionosphere, producing intermittent flashes of blue light. These effects were to be imaged by the pallet television and photometer instruments. It was hoped they would illuminate the process by which energy is transferred from the ionosphere to the atmosphere.

Another five investigations, using pallet-mounted spectrometers and a camera looking through the module's optical window, observed upper atmosphere phenomena, such as airglow, mysterious cloudlike structures, and the interaction of sunlight with hydrogen atoms.

Life science investigations using the payload specialists as subjects were focused on the physiological effects of microgravity. The most conspicuous effect, space sickness syndrome, was the target of three studies by NASA and ESA scientists. Related studies sought evidence of the effects of microgravity on human vestibular function, immune response, changes in red blood cell

mass, distribution of body fluids, cardiovascular function, lymphocyte proliferation, and perception of mass in space. A related experiment tested effects of spaceflight on circadian (biological) rhythm in plants.

In addition to the pallet cosmic ray detector, a NASA sensor measuring the intensity and energy of cosmic rays penetrating the laboratory was installed inside. Two related experiments by the German Institute for Flight Medicine sought to determine the effect of cosmic rays on biological material sandwiched between cosmic ray detectors in the module and on the pallet.

Navigation requirements for conducting some of the experiments involved 150 changes in *Columbia*'s attitude. These changes sometimes seemed to interfere with voice and data transmission through *TDRS 1*, but in general the communications were pronounced satisfactory.

At Houston, a Payload Operations Control Center (POCC) had been set up at the Johnson Space Center. It enabled investigators to follow the progress of their experiments, communicate with the crew, or send commands to the computer that controlled the experiments aboard Spacelab. The POCC parallelled the operations of Mission Control itself and received a vast flow of scientific data through its own computer. Materials science experiment data were also transmitted to the West German Aerospace Research Establishment near Munich for the convenience of European materials researchers.

As on other space missions, the crew saved the day for some experimenters when instruments malfunctioned. Only three of the 38 primary experiments were reported to have lost significant amounts of data because of failure.

Loss of data from the active cavity radiometer, one of three instruments measuring the solar constant, was averted by maneuvering *Columbia* to put a data relay device in the shade. The device shut down when heated by direct sunlight.

With instructions from Houston, Parker freed a jammed film magazine in the ESA metric camera. He rigged a darkroom in a sleeping bag on *Columbia*'s mid-deck. He was then able to open the magazine and fix it without exposing 400 frames of black-and-white film.

Parker also freed a jammed mechanism in the Spacelab high data recorder. This machine recorded vast amounts of instrument data during portions of the orbit where *Columbia* was too distant from *TDRS 1* to relay data through the satellite.

Following advice from the ground, Lichtenberg modified a fluid physics module to halt spills of liquids used in a series of microgravity tests. Merbold made three repairs to the isothermal furnace in the materials science rack.

During the mission, there were intermittent conferences between members of the crew and several heads of state. The first took place December 4 between Garriott, using a four-watt, hand-held transceiver, and a fellow amateur radio buff in Jordan as *Columbia* passed over the Middle East.

"Hello Juliet, Yankee One calling," said Garriott in ham radio speech. "This is W-Five Lima Foxtrot Lima in the spacecraft *Columbia* calling and standing by, over. . . ."

"Hello, Whiskey Five Lima Foxtrot Lima, this is Juliet," came the voice from the ground. "Good morning, sir."

"Good morning, sir, this is W-Five Lima Foxtrot Lima coming right back," said Garriott. "Your signals are five by nine [loud and clear]. I just passed over the Red Sea and the Gulf of Aqaba. I'm looking down on your country right at this time, sir. Is this your royal highness speaking? Over."

"Whiskey Five Lima Foxtrot Lima, this is Juliet, Yankee One. Hussein at the mike. Also very happy to hear from you. You are five by nine also here in the city of Amman and you have a very fine signal and we're happy to hear from you. We're very excited about this first contact with *Columbia*. . . . Go ahead, sir."

Garriott then explained that he was radiating only four watts from his little radio through a homemade antenna rigged by the amateur radio club at the Johnson Space Center, Houston.

"I've had a marvelous time using it," he said. "As a matter of fact, you are my first contact outside of North America. And perhaps you would be interested in my view . . . looking across Jordan to the Mediterranean and as is usual the sky is clear and there is a few cirrus

clouds scattered over so I can see through all of that. Fantastic view over here, looking down on Amman. Go ahead, please."

"Thank you very, very much," replied King Hussein. "We're very proud and happy indeed to send you and your colleagues from all my countrymen and myself our very, very best for a most successful mission and best wishes for a safe trip. And we're very proud of you and we share this pride with all the people of America, all the people of the world. . . ."

The next day, there was a televised conference call via *TDRS 1* among the crew, President Reagan in Washington, and West German Chancellor Helmut Kohl, who was in Athens, attending a summit meeting of the European Economic Community.

President Reagan: Chancellor Kohl is with us all the way from Athens, Greece, along with you astronauts hovering in space and with me here in Washington and the whole world listening. This is one heck of a conference call! Seriously, though, this Space Shuttle mission represents the enormous potential available to mankind. I know Chancellor Kohl agrees with me on that. . . .

Chancellor Kohl: Thank you very much, Mr. President. It is a terrific experience for us to be able to talk together this way and to talk to the crew as well. Above all, I would like to send my best wishes to my countryman, dear Herr Merbold.

Merbold: Guten tag, Bundeskanzler.

There followed a televised tour of the Spacelab module, conducted by Merbold and Lichtenberg, who described some of the experiments. On the planet below, the two heads of state voiced their enthusiasm and congratulated the crew.

"Our investment in space has been an exceptional bargain," Mr. Reagan concluded.

With STS-9–Spacelab 1, the first element of the Presidential Space Task Group Report of 1969—the ground-to-orbit transport—was fully realized. On the evening of January 25, 1984, the second element of that 15-year-old blueprint—the permanent space station—was given a long-awaited "Go" by President Reagan in his State of the Union message to Congress. In language reminiscent of President Kennedy's call in 1961 for a manned lunar landing "before this decade is out," Mr. Reagan said, "Tonight, I am directing NASA to develop a permanent space station and to do it within a decade."

NASA was ready. As Administrator Beggs explained at a news conference the next day, the Space Station Task Force planned to use the Shuttle to carry up the pieces—the modules—from which the station would be assembled in low Earth orbit. But before "any metal-bending contracts are let" for the project, Beggs said, there would be an extended period of defining the structure in the hope of avoiding the trouble NASA had experienced in building the Shuttle.

During the definition phase of the project, the station would undergo metamorphosis as the Shuttle had done, so that its size and shape could merely be approximated in early 1984. It could carry a crew of up to eight and would orbit the Earth at an altitude between 200 and 300 miles. Initial construction, with completion estimated in the early 1990s, would cost about $8 billion, but this would buy only the basic outpost on what Mr. Reagan called "our next frontier." The station would continue to develop in the twenty-first century as its scientific, commercial, and industrial operations expanded and as it became an increasingly busy staging facility for flights, manned and unmanned, to geostationary orbit, to the Moon, and to the planets.

Since 1969, the Shuttle and the space station had been inextricably linked in NASA's space planning. John Hodge, the Space Station Task Force director, called them a "matched pair." The Shuttle made the space station feasible and, as Hodge put it, the station gave the Shuttle someplace to go.

Flight	Crew	Vehicle	Prep Time at KSC (days)	Launch Date and Time	Inclination to Equator (degrees)	Launch Weight (pounds)	Cargo
STS-1	John W. Young, c Robert L. Crippen, p	*Columbia*	749	Apr. 12, 1981 7:00 A.M. EST	40.3	4,461,620	DFI ACIP
STS-2	Joe H. Engle, c Richard H. Truly, p	*Columbia*	198	Nov. 12, 1981 10:10 A.M. EST	38	4,475,943	OSTA-1 pallet IECM DFI ACIP
STS-3	Jack R. Lousma, c Charles G. Fullerton, p	*Columbia*	114	Mar. 22, 1982 11:00 A.M. EST	38	4,478,954	OSS-1 pallet IECM DFI
STS-4	Thomas K. Mattingly II, c Henry W. Hartsfield, Jr., p	*Columbia*	80	June 27, 1982 11:00 A.M. EDT	28.5	4,482,888	DOD 82-1 IECM DFI One GAS
STS-5	Vance DeV. Brand, c Robert F. Overmyer, p Joseph P. Allen IV, ms William B. Lenoir, ms	*Columbia*	120	Nov. 11, 1982 7:19 A.M. EST	28.5	4,488,599	*SBS 3*–PAM-D *Anik* C–PAM-D DFI
STS-6	Paul J. Weitz, c Karol J. Bobko, p Donald H. Peterson, ms F. Story Musgrave, ms	*Challenger*	274 (126 on pad)	Apr. 4, 1983 1:30 P.M. EST	28.5	4,489,843	*TDRS 1*–IUS Three GAS
STS-7	Robert L. Crippen, c Frederick H. Hauck, p John McC. Fabian, ms Sally K. Ride, ms Norman E. Thagard, ms	*Challenger*	64 (24 on pad)	June 18, 1983 7:33 A.M. EDT	28.45	4,486,141	*Palapa B*–PAM-D *Anik* C–PAM-D SPAS-01 OSTA-2 pallet Seven GAS
STS-8	Richard H. Truly, c Daniel C. Brandenstein, p Guion S. Bluford, ms Dale A. Gardner, ms William E. Thornton, ms	*Challenger*	62 (28 on pad)	Aug. 30, 1983 2:32 A.M. EDT	28.45	4,391,622	*INSAT 1B*–PAM-D DFI PFTA Twelve GAS
STS-9	John W. Young, c Brewster H. Shaw, Jr., p Owen K. Garriott, ms Robert A. R. Parker, ms Byron K. Lichtenberg, ps Ulf Merbold, ps	*Columbia*	1 year	Nov. 28, 1983 11:00 A.M. EST	57	4,503,095	Spacelab 1

c	commander	EAFB	Edwards Air Force Base
p	pilot	EDT	Eastern Daylight Time
ms	mission specialist	EST	Eastern Standard Time
ps	payload specialist	EVA	extravehicular activity
ACIP	Aerodynamic Coefficient Identification Package	GAS	Getaway Special
CFES	Continuous Flow Electrophoresis System	IECM	Induced Environmental Contamination Monitor
DFI	Development Flight Instrumentation	IUS	Inertial Upper Stage
DoD	Department of Defense	KSC	Kennedy Space Center

Cargo Weight (pounds)	Duration	Orbits	Distance (statute miles)	Landing Date and Location	Landing Weight (pounds)	Nominal Altitude (miles)	Mission
9,300	54 hr 20 min	36	933,394	Apr. 14, 1981 EAFB	196,500	166	First flight test
19,388	54 hr 13 min	36	931,358	Nov. 14, 1981 EAFB	204,000	157	Flight test RMS test NOSL
21,293	192 hr 4 min	129	3,355,071	Mar. 30, 1982 White Sands	207,500	147	Flight test RMS test MLR, Electrophoresis test
N/A	169 hr 9 min	112	3,055,193	July 4, 1982 EAFB	209,500	184–197	Final flight test First DoD cargo First commercial test (CFES) RMS repositioning of IECM MLR and NOSL
33,013	122 hr 14 min	81	2,109,298	Nov. 16, 1982 EAFB	204,103	184	First operational flight Deployment of SBS 3 and Anik C with PAM-Ds
46,615	120 hr 24 min	80	2,079,538	Apr. 9, 1983 EAFB	196,600	176.6	Deployment of TDRS 1 with IUS EVA by Peterson and Musgrave CFES, MLR, and NOSL
31,985	146 hr 24 min	97	2,334,576	June 24, 1983 EAFB	202,976	185	First U.S. woman in space Deployment of Palapa B and Anik C with PAM-Ds RMS release and retrieval of SPAS-01 CFES and MLR
26,163	121 hr 29 min	80	2,077,477	Sept. 4, 1983 EAFB	204,272	172.5	First night launch and landing Deployment of Insat 1B with PAM-D RMS unberthing and berthing of PFTA (heavy-payload test) Space sickness tests
33,584	247 hr 47 min	166	4,292,825	Dec. 8, 1983 EAFB	221,000	155.3	Spacelab 1 with 71 scientific investigations. First European scientist to fly in U.S. spacecraft.

MLR	Monodisperse Latex Reactor
N/A	not available
NOSL	Nighttime/Daytime Optical Survey of Lightning
OSS	Office of Space Science
OSTA	Office of Space and Terrestrial Application
PAM	Payload Assist Module
PFTA	Payload Flight Test Article
RMS	Remote Manipulator System

SBS	Satellite Business Systems
SPAS	Shuttle Pallet Satellite
STS	Space Transportation System
TDRS	Tracking Data and Relay Satellite

Notes

CHAPTER 1

1. Tsiolkovsky, K. *Exploration of the Universe with Reactive Devices.* Moscow, 1903, 1911; Paris, 1914, 1926. See also *Konstantin Tsiolkovsky,* a pamphlet issued by the Novosti Agency Publishing House, Moscow and London. The Novosti version stated: "The Earth is the cradle of reason but one cannot live in the cradle forever."

CHAPTER 2

1. Kissinger, Henry. *The White House Years.* Boston: Little, Brown, 1979, p. 223.

2. Mueller, George E. Opening remarks, Space Shuttle Symposium, Smithsonian Museum of Natural History, Washington, D.C., October 16–17, 1969.

3. Hunter, Maxwell W. Lockheed Aircraft Corporation Report. Space Shuttle Symposium.

4. Tischler, A. O. Introductory remarks, Space Shuttle Symposium.

5. Boland, E. P. "Where Do We Go from Here?" paper delivered at the annual meeting, American Association for the Advancement of Science, Washington, D.C., February 14, 1978.

6. Van Allen, J. A. "U.S. Space Science & Technology." *Science,* October 30, 1981.

7. Interview conducted February 12, 1977, in Washington, D.C.

8. Status Report on the Space Shuttle and Skylab. House Committee on Science and Astronautics, 93d Cong., 1st sess., January 1973.

9. Hearings, Subcommittee on NASA Oversight, House Committee on Science and Astronautics, 92d Cong., 2d sess., January 1972.

CHAPTER 3

1. *Chicago Tribune,* January 16, 1972, p. 12.

2. Congressional Record, 92d Cong., 1st sess., June 4, 1971, S8246.

3. NASA Facts, report issued October 1972 by NASA Headquarters.

4. Statement of Ralph E. Lapp submitted to the Senate Committee on Aeronautics and Space Science, 92d Cong., 2d sess., April 12, 1972.

5. Rebuttal by Klaus P. Heiss, Mathematica, Inc., before the

Senate Committee on Aeronautics and Space Science, April 12, 1972.

6. NASA authorization hearings, Senate Committee on Aeronautics and Space Science, 93d Cong., 1st sess., March 6, 1973.

7. Hearings, Subcommittee on NASA Oversight, House Committee on Science and Astronautics, 93d Cong., 2d sess., February 9–14, 1974.

CHAPTER 4

1. Carr, M. H. "Geology of Mars." *American Scientist* 68(6):626–635.

CHAPTER 5

1. NASA authorization hearings, Subcommittee on Manned Spaceflight, House Committee on Science and Astronautics, 92d Cong., 1st sess., March 1971.

2. Technical Status of the Space Shuttle Main Engine. Report of the Ad Hoc Committee for Review of the Space Shuttle Main Engine Program, March 1978. National Research Council.

3. NASA authorization hearings. Subcommittee on Science, Technology, and Space, Senate Committee on Commerce, Science, and Transportation, 96th Cong., 1st sess., May–June 1979.

CHAPTER 7

1. NASA authorization hearings, Subcommittee on Science, Technology, and Space, Senate Committee on Commerce, Science, and Transportation, 96th Cong., 1st sess., February 1979.

2. NASA authorization hearings, Subcommittee on Space Science and Applications, House Committee on Science and Technology, and Subcommittee on Science, Technology & Space, Committee on Commerce, Science, and Transportation, 96th Cong., 1st sess., April–June 1979.

CHAPTER 8

1. Rifkin, Jeremy and Howard, Ted. *Entropy: A New World View*. New York: Viking Press, 1980.

2. NASA Authorization Hearings, Subcommittee on Space Science and Applications, House Committee on Science and Technology, 95th Cong., 2d sess., January 1978.

3. Glaser, Peter E. and Burke, James C. "New Directions for Solar Energy Application," paper published by Arthur D. Little,

Inc., Cambridge, Mass., 1978, and statement by Glaser at hearings on HR 2335, The Solar Power Satellite Research, Development, and Evaluation Act of 1979, before the Subcommittee on Space Science and Applications, House Committee on Science and Technology, held March 28–30, 1979.

4. Heiss, Klaus P. "Economic Opportunities of Space Enterprise in the Next Decades," paper presented to the 16th Space Congress of the Canaveral Council of Technical Societies, Cocoa Beach, Fla., 1979.

5. Glaser, Peter E. "The Solar Power Satellite," paper delivered at the 16th Space Congress.

6. "Solar Power Satellites," an AIAA Position Paper. Nov. 27, 1978.

7. Heiss, Klaus P. "Economic Opportunities of Space Enterprise."

8. Waltz, D. M. "Prospects for Space Manufacturing," paper presented at the Bicentennial Symposium, AAS and AIAA, at the National Air and Space Museum, Washington, D.C., October 6–8, 1976. Published in *Technology Review*, May 1977.

9. "A TVA for Space," paper presented to the American Association for the Advancement of Science annual meeting, Washington, D.C., February 13, 1978.

10. "Next Steps for Mankind," a symposium held July 19, 1979. Washington, D.C.: U.S. Government Printing Office, 1980.

11. "Next Steps for Mankind," p. 32.

12. "Next Steps for Mankind," pp. 46–47.

CHAPTER 11

1. Panel discussion, STS-1 Flight Review, conducted at the 18th Space Congress of the Canaveral Council of Technical Societies, April 29, 1981.

2. "Subsurface Valleys and Geoarcheology of the Eastern Sahara Revealed by Shuttle Radar." *Science*, December 3, 1982. Authors are J. F. McCauley, G. G. Schaber, C. S. Breed, and M. J. Grolier, U.S. Geological Survey; C. V. Haynes, University of Arizona; B. Issawi, Egyptian Geological Survey; and C. Elachi and R. Blom, Jet Propulsion Laboratory.

CHAPTER 13

1. NASA authorization hearings, Subcommittee on Space Science and Applications, House Committee on Science and Technology, 97th Cong., 1st sess., March 1981.

2. Hodge, John D. "A Space Station for America." Address to AIAA/NASA Symposium on the Space Station, July 18–20, 1983, Arlington, Va.

INDEX

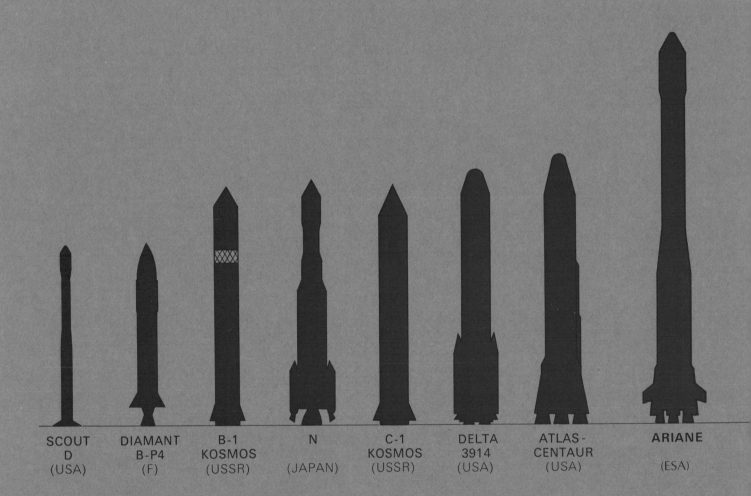

SCOUT
D
(USA)

DIAMANT
B-P4
(F)

B-1
KOSMOS
(USSR)

N
(JAPAN)

C-1
KOSMOS
(USSR)

DELTA
3914
(USA)

ATLAS-
CENTAUR
(USA)

ARIANE
(ESA)